MW01155385

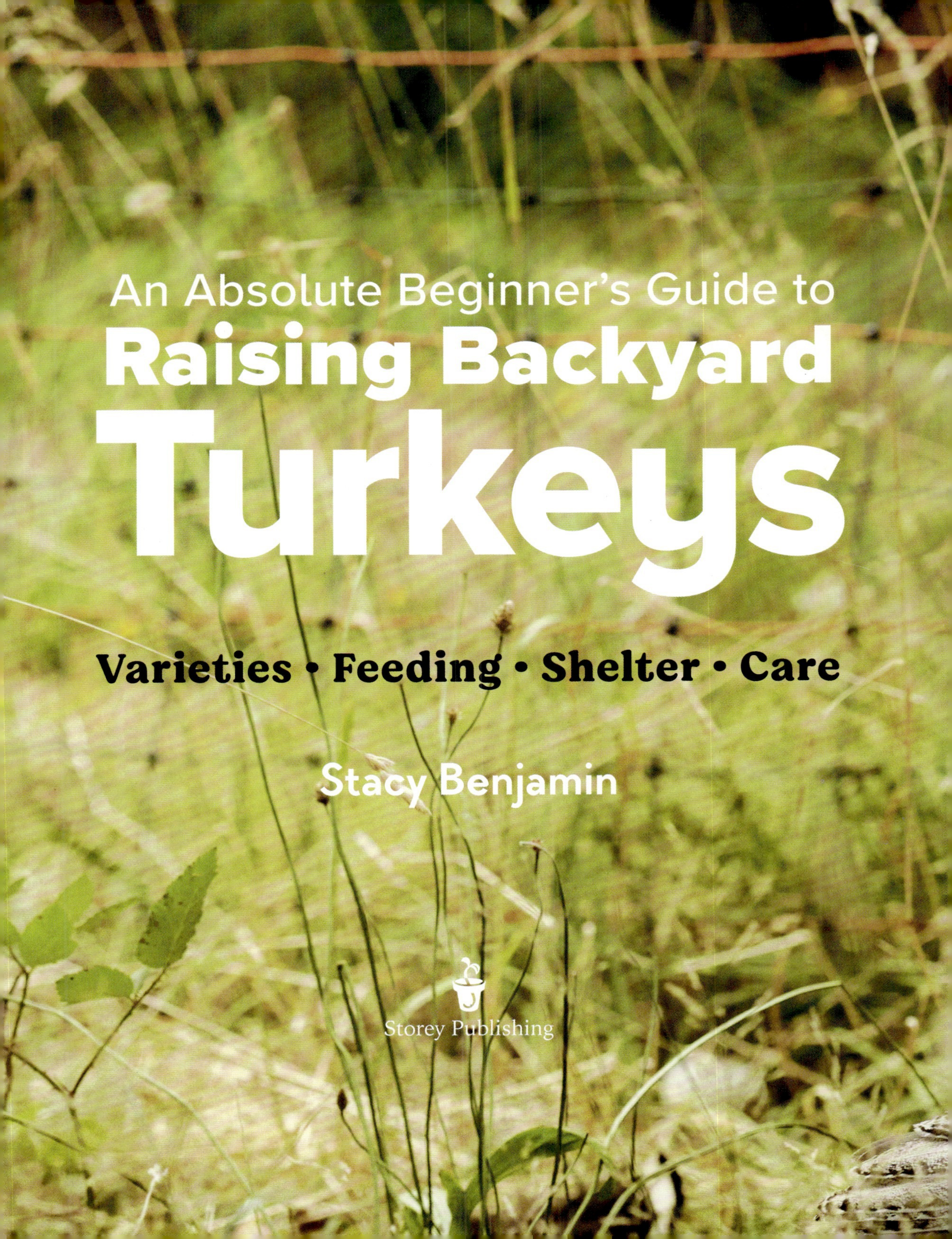

An Absolute Beginner's Guide to

Raising Backyard Turkeys

Varieties • Feeding • Shelter • Care

Stacy Benjamin

Storey Publishing

The mission of Storey Publishing is to serve our customers by publishing practical information that encourages personal independence in harmony with the environment.

Edited by Carleen Madigan and Lisa H. Hiley
Art direction and book design by Erin Dawson
Text production by Jennifer Jepson Smith

Cover photography by © Katie Newburn, front (r. all but t.); © JKemperPhotographyLLC/Shutterstock.com, front (l.); © rhammbone/iStock, front (t.r.); © Rehlik/iStock.com, back
Interior photography by © Katie Newburn
Additional photography by © Amy israel/Shutterstock.com, 23 2nd from t.; © Amy McNabb/Shutterstock.com, 22 2nd from b.; Bernardino de Sahagún/Public domain/Wikimedia Commons, 14; © chloe7992/Shutterstock.com, 22 t.; © Denys Prokofyev/Alamy Stock Photo, 94 l.; © Evoken/Shutterstock.com, 23 t.; Frank Reese courtesy of The Livestock Conservancy, 23 b.; © goldfinch4ever/123RF, 62; © Jared S Davies/Shutterstock.com, 16 r.; © Jennifer Yakey-Ault/Shutterstock.com, 16 l.; © Lepas/Shutterstock.com, 94 r.; © Marynag/Dreamstime.com, 61; © Michael Chatt/Shutterstock.com, 59; © Pukhov K/Shutterstock.com, 140; © slowmotiongli/Shutterstock.com, 23 2nd from b.; © Smiler99/Shutterstock.com, 22 2nd from t.; © Stacy Benjamin, 5, 11, 27, 30, 37, 38, 40, 45 l., 50, 52, 54, 63, 68, 71, 90, 91, 96, 97, 100, 101, 105 b., 107, 111–113, 119, 120, 124, 126, 127 b., 129, 130; © Stephen Gibson/Shutterstock.com, 15; © Tim Belyk/Shutterstock.com, 18
Illustrations by © Elayne Sears

Text © 2025 by Stacy Benjamin

All rights reserved. Hachette Book Group supports the right to free expression and the value of copyright. The purpose of copyright is to encourage writers and artists to produce the creative works that enrich our culture. The scanning, uploading, and distribution of this book without permission is a theft of the author's intellectual property. If you would like permission to use material from the book (other than for review purposes), please contact permissions@hbgusa.com. Thank you for your support of the author's rights.

The information in this book is true and complete to the best of our knowledge. All recommendations are made without guarantee on the part of the author or Storey Publishing. The author and publisher disclaim any liability in connection with the use of this information.

The publisher is not responsible for websites (or their content) that are not owned by the publisher.

Storey books may be purchased in bulk for business, educational, or promotional use. Special editions or book excerpts can also be created to specification. For details, please contact your local bookseller or the Hachette Book Group Special Markets Department at special.markets@hbgusa.com.

Storey Publishing
210 MASS MoCA Way
North Adams, MA 01247
storey.com

Storey Publishing is an imprint of Workman Publishing, a division of Hachette Book Group, Inc., 1290 Avenue of the Americas, New York, NY 10104. The Storey Publishing name and logo are registered trademarks of Hachette Book Group, Inc.

ISBNs: 978-1-63586-756-5 (paperback);
978-1-63586-757-2 (ebook)

Printed in China through World Print on paper from responsible sources
10 9 8 7 6 5 4 3 2 1

Library of Congress Cataloging-in-Publication Data on file

This book is dedicated to my husband, Sean, for his devotion to 5R Farm, for making my dreams come true, and for working tirelessly to keep the farm running smoothly and looking good. For the many projects he has undertaken to give our turkeys the best life possible—whether they were used as intended or not at all—and for the many pecks he endured from my sassy tom turkey Ringo, I want to express my deepest gratitude on behalf of unappreciative turkeys everywhere.

Contents

PREFACE
For the Love of Turkeys
9

CHAPTER 1
Why Raise Turkeys?
13

CHAPTER 2
What to Expect When Keeping Turkeys
33

CHAPTER 3
Turkey Behavior
43

CHAPTER 4
Shelter and Space Requirements
65

CHAPTER 5
Nutrition and Care
79

CHAPTER 6
Raising Poults from Day 1
93

CHAPTER 7
Natural Brooding and Raising
115

CHAPTER 8
Healthcare
133

Glossary 144
Resources 146
Acknowledgments 147
Index 148

For the Love of Turkeys

I HAD NO IDEA WHEN I GOT MY FIRST three backyard chickens that they were the beginning of my journey toward the wonderful life with turkeys that I enjoy today. After college and a brief stint in New York City, my husband, Sean, and I bought a house in Portland, Oregon. As the backyard chicken movement swept the city, I began to hear the happy sounds of chickens as I walked through the neighborhood. I started researching all I could find about keeping backyard chickens, and I soon fell in love with them through the beautiful photos I saw online and the tales I read of their quirky behaviors and personalities. I thought they sounded like the perfect little companions for an animal lover like me.

After several months discussing with my husband the pros and cons of getting a flock of chickens, I finally convinced him we needed to have them. They are, after all, the pet that makes you breakfast. Sean built our first of many chicken coops, which was dubbed "The Poultry Palace" by our neighbor, and we welcomed our first three chicks in the spring of 2010: Rhoda the Rhode Island Red, Raquel the Barred Plymouth Rock, and an Easter Egger that turned out to be a rooster and had to be rehomed because we lived in the city. At that point, I obviously had no choice but to add three more chicks named Rosie, Ruby, and Ramona.

I was completely enamored with my feathered ladies, and I soon began to wish I had room for more chickens. For several years we had dreamed about buying a place in the country, but after we got chickens the conversation became more serious. Within a year of acquiring our first chickens, we were the proud owners of 4.5 acres in St. Helens, Oregon. We named the property 5R Farm, after the five chickens who inspired us to make our dream of moving to the country a reality.

It didn't take long for me to begin considering the other types of livestock I might add to the farm. I found a blog that said if you enjoyed raising chickens, you would absolutely love raising turkeys, which planted the seed in my mind to get turkeys. After researching the differences between heritage turkeys and broad breasted turkeys, I decided on heritage turkeys and then settled on the Narragansett variety, because I wanted to keep a self-sustaining flock and I read that they made good mothers. I've been so happy with my decision to raise Narragansetts.

One of my favorite things to do is to share the joys of turkey ownership with others and encourage them to add these wonderful birds to their homestead. I imagine many of the people reading this book will have some experience with raising poultry, likely with chickens, which are well known as gateway farm animals. Consider everything that you adore about your chickens—their beautiful feathers, charismatic ways, cute sounds, and different personalities—and then multiply that adoration by a factor of 10 as preparation for life as a turkey owner.

If you have experience raising chickens, that will be helpful as you begin your journey raising turkeys. Make no mistake, it will be a journey—one that is filled with lots of happy surprises along the way, as well as a fair share of problem-solving. I hope to give you an understanding of what to expect before you jump into the wonderful world of turkey ownership, as well as helpful information that will make your experience of raising turkeys as enjoyable as mine has been.

FROM FOOD TO FARM FRIEND

When we decided to add heritage turkeys to our farm, the initial plan was to keep a breeding trio year-round and hatch a few poults in the spring to provide turkey dinners for the holidays. I already knew that I became too easily attached to my chickens to be able to raise them as a source of meat, even though that had been one of the goals with our first flock of heritage chickens. Nevertheless, we returned to the idea of raising our own meat birds when I started to research turkeys.

Before bringing them home, I thought, “Tom turkeys are kinda funny looking, so surely that will make it easier to remember that they are intended to be dinner and not to think of them as pets.” But those funny-looking and oh-so-charismatic turkeys made quick work of that plan. Over the first couple of years we harvested a few, but every year the size of our flock grew as I decided to keep new favorites from the spring hatch.

As I write this, we have a flock of eight adult turkeys—two toms and six hens. We have had as many as 20 adults and juveniles after a spring hatch. We now keep turkeys mainly for eggs and entertainment, and I have accepted that I just don’t have it in me to raise and harvest birds for the table. As much as I respect the idea and admire those who can do it year after year, my soft heart gets in the way. There is so much to love about turkeys, whether their intended purpose is as food or friend, and I can’t imagine life on the farm without them.

Ringo and Pumpkin Pie are just two of the charismatic turkeys who earned a forever home on our farm.

CHAPTER 1

Why Raise Turkeys?

The most common answer to this question is to raise your own turkey for Thanksgiving or Christmas dinner. Or perhaps you'd like to fill your freezer with a lean source of protein to enjoy year-round. There is nothing more satisfying than providing your family with nutritious food that you've raised or grown in your own backyard. In addition to providing meat for the table, heritage turkeys are a productive source of extra-large eggs.

Even if you don't eat meat or need a source of eggs, the entertainment value alone is a good reason to add a few turkeys to your homestead. One of my favorite sights is a tom turkey in full strut, gobbling a welcome to our farm guests. Everyone who visits our farm is instantly drawn to the turkeys in the front yard. The turkeys come running to investigate the visitors, and our guests hurry to get out their cameras as they ooh and aah over our beautiful birds.

Turkeys also make excellent guardians for other types of poultry because they keep a watchful eye over their surroundings and are quick to sound an alarm call when they notice potential predators, whether they be in the sky or on the ground.

A Brief History of Domestic Turkeys

HERITAGE TURKEYS ARE DESCENDED FROM the wild turkeys (*Meleagris gallopavo*) that are native to the Americas. That the birds of today retain many of their ancestors' wild behaviors and instincts is one of the things I find so fascinating about them. Although broad-breasted turkeys also developed from wild turkeys, many of those ancestral characteristics have been bred out of them.

Heritage Turkeys Came First

Wild turkeys were domesticated by the Indigenous peoples of the southwestern United States and parts of Mexico more than 2,000 years ago. Early explorers of the Americas brought some of these domesticated turkeys back to Europe, and over time the birds spread across the European continent. British settlers later brought domesticated turkeys back to North America when they colonized New England in the 1600s.

Turkeys were common on family farms in the eighteenth and nineteenth centuries. The American heritage turkey varieties we know today are descended from crosses of European varieties with eastern wild turkeys. Two of the earliest domesticated varieties were the Black and the Bronze. New variations arose along the East Coast as breeders began selecting for traits such as color and performance. Selection for distinctive color patterns resulted in the development of striking breeds such as the Narragansett and Royal Palm.

Broad Breasted Turkeys and Factory Farming

The Broad Breasted Bronze turkey shares the origin of heritage turkeys, but development diverged sharply from that of heritage turkeys with the onset of factory poultry farming in the 1920s. The public's desire for bigger birds with larger breasts combined with poultry producers' desire to raise bigger birds quickly to market weight resulted in the development of the Broad Breasted Bronze.

The dark feathers of the Broad Breasted Bronze meant that dark pinfeathers were sometimes visible in the dressed carcass, giving it an undesirable appearance. To solve this problem, producers developed the Broad Breasted White, a cross between the Broad Breasted Bronze and the White Holland heritage turkey. White feathers produce a cleaner-looking carcass, and by the 1960s this was the dominant turkey on commercial poultry farms.

Broad-breasted turkeys can potentially live for several years, although it is common for them to develop adverse health effects (joint issues, lameness, heart or kidney failure) as they age and continue to grow

Broad-breasted turkeys were developed to grow quickly so they could be processed for market within a few months.

Comparing Heritage and Broad Breasted Turkeys

Heritage turkeys take longer to reach a good harvest weight and are on average about half the size of broad breasted turkeys at maturity. Heritage turkeys can reproduce without artificial insemination, and the hens are typically attentive mothers. Heritage turkeys require relatively minimal care to live long, healthy lives; broad breasted varieties are not bred to live beyond their first year.

BROAD BREASTED VARIETIES

- Can weigh up to 50 pounds
- Reach harvest weight in 4 to 5 months
- Excellent feed conversion ratio
- Cannot reproduce without artificial insemination
- Shorter lifespan due to weight-related health issues

HERITAGE VARIETIES

- Generally weigh 20 to 25 pounds
- Reach harvest weight in 6 to 7 months
- More expensive to raise to harvest weight
- Can reproduce naturally
- Can live up to 12 years

larger. These problems are rare in the four- to six-month lifespan of a broad-breasted turkey raised for the table, but health issues occur in approximately 50 percent of birds that live longer than six months. Many don't live beyond two years.

I've heard time and time again from heartbroken turkey owners who decided to make their broad-breasted turkey a pet instead of dinner. Fast-forward a few months past Thanksgiving and their beloved pet begins to suffer serious health problems or is even unable to walk. For this reason alone, I would advise against getting broad-breasted turkeys. Why not get a heritage variety instead and give yourself more options for the future?

What Kind of Turkey Is Right for You?

CHOOSING WHETHER TO KEEP heritage or broad-breasted turkeys is a major factor in how you will raise your birds, one that will define your relationship with them. For many decades, broad-breasted turkeys were the most common type of turkey raised on both small and large farms. In recent years, however, heritage turkeys have experienced a resurgence in popularity among farmers and hobbyists who are interested in preserving a bit of early American agricultural history.

Turkeys make wonderful lawn ornaments!

Raising Broad-Breasted Turkeys

For many people, the most familiar image of a turkey is the broad-breasted mainstay of the typical Thanksgiving dinner table. Broad-breasted turkeys were developed by the poultry industry with the goal of creating a rapidly growing bird with lots of white breast meat that could be economically raised for the Thanksgiving market. Broad-breasted turkeys have an excellent feed conversion ratio, meaning they efficiently process food into growth, which in turn means they reach market weight in just four to five months. Toms can grow to 50 pounds, although they are often harvested before they reach maturity.

The better feed conversion ratio of broad-breasted turkeys may be a consideration for those looking to raise turkeys as a profitable endeavor. But for those who want to raise a few turkeys for themselves, the

Having bronze varieties of both heritage and broad-breasted turkeys can lead to confusion. Not all bronze turkeys are considered heritage birds. The term "Standard Bronze" differentiates the Bronze heritage turkey variety from the Broad Breasted Bronze turkey shown here.

difference in feed conversion between broad-breasted and heritage turkeys is not significant enough to discourage one from raising heritage varieties.

Broad-breasted turkeys cannot reproduce naturally because they have been artificially selected to have abnormally large breasts, which hinders mating. The commercial turkey industry uses artificial insemination to produce each generation of Thanksgiving turkeys. As a result of their disproportionate size and shape, broad-breasted turkeys typically experience numerous adverse health effects if they are kept beyond their intended short lifespan. For that reason, they are not a good choice if your goal is to keep a breeding flock of turkeys.

COLOR VARIETIES. Broad-breasted turkeys are available in only two colors: bronze and white. This can lead to confusion when purchasing poults (young turkeys) at the feed store because there are also bronze and white varieties of heritage turkeys. Because broad-breasted varieties still dominate the turkey market, you should assume that turkey poults marked simply "bronze turkey" or "white turkey" are the broad-breasted varieties.

Making the Case for Heritage Turkeys

If your intention is to keep a self-sustaining and naturally reproducing flock, or rafter, then heritage turkeys are the way to go. Heritage turkeys tend to be healthy, hardy birds that can live for 10 to 12 years. Even if you plan to raise only a couple of turkeys for holiday meals, I encourage you to get a heritage variety. One consideration is flavor. Many people's palates have become accustomed to the relatively flavorless breast meat of a broad-breasted turkey. The meat of a heritage turkey is juicier and more flavorful, in the same way that pasture-raised beef is more flavorful than feedlot beef.

Heritage turkeys make a beautiful year-round addition to the homestead or small farm.

Be forewarned, though: I know quite a few people who intended to raise a Thanksgiving turkey only to become so enamored with these beautiful and charismatic birds that they pardoned their intended dinner and gained a beloved farm friend.

Heritage turkeys are defined by the following characteristics, according to The Livestock Conservancy.

- Capable of reproducing through natural mating
- Slow growth rate
- Long, productive outdoor lifespan

A truly wonderful part of the turkey experience is seeing how well-adapted turkeys are to their natural environment through the seasons—nesting and laying eggs in the tall grass in the spring, raising little ones in the summer, and withstanding cold winters and snow like champs.

Heritage turkeys make a beautiful year-round addition to the homestead or small farm. If you already have chickens and you are looking to add a little more visual interest to your flock, heritage turkeys will fit that role splendidly. Or maybe you don't have chickens because they are too much daily work, and you would like to keep a hardier, more independent bird that can practically take care of itself. In that case, heritage turkeys may be just what you are looking for.

One of the main reasons I enjoy having turkeys is the beautiful eggs the hens lay from spring through fall. Turkey eggshells are a light cream to light tan with the most beautiful brown speckles. If you are a fan of colored chicken eggs, then having turkeys in your flock will increase the "eye candy" appeal of your egg basket tremendously.

Turkeys lay beautifully speckled eggs.

It's hard to resist the companionship of a charismatic tom turkey.

HERITAGE VARIETIES. The different types of heritage turkeys are known as varieties, not breeds. The following varieties are accepted by the American Poultry Association (APA) for inclusion in its Standard of Perfection.

- Beltsville Small White
- Black
- Bourbon Red
- Narragansett
- Royal Palm
- Slate
- Standard Bronze
- White Holland

Many other colorful varieties of heritage turkeys that are not recognized by the APA are available from hatcheries and hobby breeders. These include Chocolate, Jersey Buff, Mottled, and Sweetgrass, to name a few, and the list goes on as new color varieties are continually being developed.

Characteristics and Origin of Heritage Turkey Varieties

BELTSVILLE SMALL WHITE

MATURE WEIGHT (TOM/HEN): Up to 21/17 pounds

COLOR: White

ORIGIN: Created in the 1940s at the University of Massachusetts to meet anticipated consumer demand for a small, meaty turkey, this variety was nearly extinct by the 1970s, after the demand failed to materialize. It remains quite rare.

BLACK/NORFOLK BLACK/BLACK SPANISH

MATURE WEIGHT (TOM/HEN): Up to 27/18 pounds

COLOR: Black with an iridescent green sheen

ORIGIN: The black turkey originated in Europe as a descendant of turkeys from Mexico that were brought to Europe by Spanish explorers. The Black variety of today was developed in the Americas after it returned with early settlers who further crossed it with native wild turkeys.

BOURBON RED

MATURE WEIGHT (TOM/HEN): Up to 32/18 pounds

COLOR: Chestnut to dark red body with white flight feathers; white tail feathers are tipped with lighter red

ORIGIN: The Bourbon Red turkey is named for Bourbon County in Kentucky, where it originated in the late 1800s. It was developed from crosses between Buff, Bronze, and White Holland turkeys.

NARRAGANSETT

MATURE WEIGHT (TOM/HEN): Up to 28/18 pounds

COLOR: Alternating bands of black/gray and white on the body and wings, with a black saddle; barred tail, mostly tan and black, with a white tip

ORIGIN: Named for Narragansett Bay in Rhode Island, where it was developed, this variety is a cross between domestic turkeys, possibly Norfolk Blacks, and wild turkeys. It was once an important part of the turkey industry in New England.

ROYAL PALM

MATURE WEIGHT (TOM/HEN): 22/12 pounds

COLOR: White with a black saddle, black feather tips, and a black band near the tip of the tail feathers

ORIGIN: One of the smaller heritage turkeys, the Royal Palm is a striking bird with alternating bands of white and black feathers. This unusual pattern first appeared in the 1920s in a mixed flock of Black, Bronze, Narragansett, and wild turkeys.

SLATE

MATURE WEIGHT (TOM/HEN): 33/18 pounds

COLOR: Blue to gray-blue, with or without black flecks

ORIGIN: There is some uncertainty about the origin of this variety; the ashy blue coloration is the result of a genetic mutation. Different mutations produce variations of this color: blue, black, or blue flecked with black spots.

STANDARD BRONZE

MATURE WEIGHT (TOM/HEN): Up to 38/22 pounds

COLOR: Dark brown with an iridescent copper sheen; white bars on the wing feathers; white-and-black bands on the tips of the tail feathers

ORIGIN: Traditionally one of the most popular varieties of heritage turkey in North America, this is one of the earliest varieties of domesticated turkeys. It was recognized by the American Poultry Association in 1874.

WHITE HOLLAND

MATURE WEIGHT (TOM/HEN): Up to 33/18 pounds

COLOR: White

ORIGIN: Although the name implies a Dutch origin, this is an American variety developed from white mutations of the Bronze turkey. It is quite rare today.

The Importance of Conservation

Heritage livestock breeds contain important genetic diversity and are a valuable asset in ensuring our future food security.

ONE OF THE MOST APPEALING CONSIDERATIONS for me in deciding to raise heritage turkeys was that I could contribute to the effort to preserve and perpetuate rare livestock animals. In this age of factory-farmed animals that have been artificially selected to maximize profits, many traditional livestock breeds that were common only a few generations ago are facing extinction. Heritage livestock breeds are adapted to their natural environment and have traits that bolster their ability to forage, reproduce, and survive in that environment. They contain important genetic diversity and are a valuable asset in ensuring our future food security in an ever-changing climate.

The Livestock Conservancy is a nonprofit organization established with the mission "to protect America's endangered livestock and poultry breeds from extinction." The Livestock Conservancy maintains a conservation priority list of rare breeds of livestock, including turkeys. The categories of rare breeds are defined on the organization's website as follows.

> **CRITICAL:** Breeds with fewer than 200 annual registrations in the United States and an estimated global population of less than 500. For poultry, fewer than 500 birds in the US, with five or fewer primary breeding flocks (50 birds or more), and an estimated global population of less than 1,000.
>
> **THREATENED:** Breeds with fewer than 1,000 annual registrations in the United States and an estimated global population of less than 5,000. For poultry, fewer than 1,000 breeding birds in the US, with seven or fewer primary breeding flocks, and an estimated global population of less than 5,000.
>
> **WATCH:** Breeds that present genetic or numerical concerns or have a limited geographic distribution, with fewer than 2,500 annual registrations in the United States, and an estimated global population of less than 10,000. For poultry, fewer than 5,000 breeding birds in the US, with 10 or fewer primary breeding flocks, and an estimated global population of less than 10,000.

In 1997, The Livestock Conservancy found only 1,335 heritage breeding turkeys in the US. With Conservancy member support, that number increased by 750 percent over a decade. The 2023 conservation priority list for turkeys classifies Beltsville Small White and White Holland as Threatened. Black, Bourbon Red, Narragansett, Royal Palm, Slate, and Standard Bronze are on the Watch list.

If you are positive that you want to raise turkeys for the table in the shortest amount of time possible and you aren't swayed by the appeal of heritage turkeys, then broad-breasted turkeys may be the right type of turkey for you. However, if you are interested in experiencing everything there is to know and love about the fascinating lives of turkeys, I encourage you to start with heritage turkeys. This book deals mainly with my experience raising heritage Narragansett turkeys. There are many similarities but also some important differences in raising heritage versus broad-breasted turkeys. Where there are differences between the two types, I'll address them.

Temperament and Behavior

TURKEY POULTS ARE VERY CURIOUS, which makes them especially endearing, and you will likely find yourself becoming attached to them in just a few days, regardless of your initial intentions for raising turkeys. Whereas young chicks are often skittish and may run away when you try to handle them, turkey poults tend to have the opposite reaction—they will often come up to your hand to investigate. The soulful eyes and inquisitive nature of young turkeys will make you fall for them in a hurry.

Poults are lively and curious, always eager to explore their surroundings.

As they grow up, turkeys become increasingly independent and eager to explore their surroundings. It's fun to watch them thoroughly investigate every new item they encounter and to listen to their vocalizations as they seem to exclaim "What is this?" over and over in the most adorable way.

Adult broad-breasted turkeys have a reputation for being more docile and friendlier toward people than adult heritage turkeys. They are generally willing to be approached and will allow

A poult that is handled gently and fed by hand is likely to become a friendly, inquisitive bird that enjoys human interaction.

themselves to be petted or touched, whereas it may take a little work to get to this level of interaction with heritage turkeys. Heritage turkeys will often readily come within a few feet of you, but they may not be as accepting of being touched or held as a broad-breasted turkey.

The likelihood of adult turkeys interacting with their owners has much to do with how they are raised and how much socializing they receive. Just as with chickens, turkeys raised by hand from a young age tend to be more interested in approaching and interacting with you when they are adults. In fact, a turkey that you've raised from a days-old poult will likely follow you around like a big, feathered puppy. Given free range of your homestead, it may try to take up residence on your porch or at least peer in the window looking for treats!

If you let a turkey hen raise the poults or if you acquire partially grown turkeys, you can expect them to be more skittish and wary.

Can Turkeys Live with Chickens?

THIS IS A COMMON QUESTION for first-time turkey owners, and it's an important one. The answer depends on whether blackhead disease (histomoniasis) is present in the region where you live. Turkeys can contract blackhead disease by foraging in areas where infected bird droppings are present or through bird-to-bird transmission. The disease is rarely fatal in chickens, which tend to be asymptomatic carriers, but chickens can easily spread the disease to turkeys, which often succumb to it (see Chapter 8). Contact your local agricultural extension office to find out whether blackhead disease is present in your area.

If it is, your turkeys should not share any ground that chickens have used over the past few years, and turkeys should be housed separately from chickens. If it's not, as is the case in many parts of the country, it's perfectly fine to let turkeys share a pasture and housing with chickens.

Do Turkeys and Chickens Get Along?

Turkeys can coexist in the same yard as chickens, but there are a few things to keep in mind to facilitate success. Turkeys can be territorial, especially at feeding time, and they may chase chickens away from the food. Turkeys also occasionally bully chickens that have docile temperaments.

I would not keep chickens with big feather crests that may block their vision in the same living area as turkeys, and I also would not keep some of the smaller and shyer breeds of chickens with turkeys. That's not to say that they couldn't coexist, but it may not be the ideal situation for these types of chickens. Providing as much space for your flock as you can and including multiple small shelters, perches, and visual screens to provide areas for chickens to escape from bullying by turkeys is always a good idea.

Fights between turkeys and roosters may occur, especially in the spring when everyone's hormones are on the uptick. Typically, it is teenage male turkeys and turkey hens that seem to take the most issue with roosters. A young rooster attempting to romance a turkey hen will get a quick lesson from both the turkey hen and her potential male turkey suitors that this is not acceptable behavior. Due to their larger size, turkeys

Turkeys and chickens can be raised together, provided that blackhead disease is not present in your geographic area.

Turkeys and chickens can coexist peacefully, given a large enough area.

typically end up victorious in these confrontations. Roosters can be notoriously stubborn, so keep an eye out for any dominance disputes that don't seem to be resolving themselves because you may need to move the rooster to another living space for his own safety.

Turkeys and chickens generally tend to stay in their own species groups even if they live in the same yard. Feeding time, however, brings everyone together, and turkeys tend to get excited when food is involved. Turkeys will proactively enforce the pecking order, and unlike roosters, tom turkeys do not stand by patiently allowing others to eat first. Tom turkeys will dive in first at feeding and treat time, so it's a good idea to have multiple feeding stations and spread out the feed and treats to allow the less dominant and/or smaller members of the flock to maneuver around the outskirts of the feeding frenzy and find a less competitive area to eat.

The pecking order is a mysterious thing, and I've had a few surprises in my mixed turkey and chicken flock. I once had a few very bossy Black Copper Marans chickens that would chase my tom turkeys away at treat time. I also had a tiny Silkie rooster that liked to go after my dominant tom turkey, which was a hilarious sight! You can't always predict how the pecking order will shake out, so give it a try with a mixed flock and be prepared with a backup plan.

Benefits of Keeping Turkeys with Chickens

Both male and female turkeys make great guardians in mixed flocks. They have excellent vision, and toms will often gobble at the sight of an aerial predator flying so high in the sky that it's difficult for us to see. Should the predator get closer, the hens will chime in, and this gives plenty of time for everyone to run to safety or take cover. The chickens seem to be attuned to the meaning of the warning gobbles and alarm calls, and I have seen raptors—even bald eagles—attempt to attack the chickens in my turkey yard, only to be thwarted due to the early warning calls from the turkeys.

Risks of Keeping Turkeys with Chickens

For the most part, my turkeys and chickens coexist well, although there were two sad occasions when I lost chickens due to the actions of my turkeys. The first incident was when I found one of my Speckled Sussex hens dead and mysteriously skinned down one side. My first thought was of an aerial predator, but there were no plucked feathers in the yard and no other signs of injury other than the skinning. I knew she was a subservient hen and would squat for practically anyone. When I looked at the spur of my tom turkey and saw that the tip was bloody, I knew immediately what had happened. It's a good idea to observe how your tom behaves around your chickens and make certain that he isn't paying too much attention to any one chicken, as he might try to mount her and in so doing injure or kill her.

The second incident was when I had an aging hen that was slowing down and had been chased by the turkey hens a few times. One day I came out to find her dead behind the coop with a bloody comb and her eyes pecked out. It was not the dignified ending I had hoped for this senior lady. So again, keep an eye on how your turkeys interact with your chickens, and be aware that you may need to find safer accommodations for a chicken that is receiving too much attention from the turkeys. As much as I love them, I'll be the first to tell you that turkeys can sometimes disappoint you with their aggressive behavior.

Turkeys can be territorial, especially at feeding time, and they may chase chickens away from the food.

Although they have different housing needs, turkeys and chickens can share a yard during the day.

TURKEY TRIBE

When we brought home our two turkey hens, Eleanor and Prudence, their only company in the pasture was our tom turkey, Ringo; a rooster named Ramon; and Ramon's three hens. The chickens had been living in the pasture on their own for several months before we introduced the turkeys, and Ramon's ladies had taken ownership of the pasture. When we added Eleanor and Prudence, Ramon's ladies let the turkey hens know in no uncertain terms that the chickens were in charge and got first dibs at treat time.

Eleanor and Prudence were a bit skittish, and I was trying to train them to come when called and to eat treats out of my hand. Every time Eleanor and Prudence would approach me for treats, one of Ramon's girls would run up and peck them or chase them away. As I watched the interactions of the turkey hens and the chickens from a distance, I saw that Prudence routinely would turn around and run the other way when one of Ramon's hens approached her. Eleanor, the bolder turkey, would sometimes chase after Ramon's girls when they went after Prudence. This went on for several months, until one day I noticed that instead of running away from Ramon's ladies, Prudence was seeking out their company.

Before I got turkeys, I did my research and read that when turkeys and chickens are kept together, they will typically hang out within their own species groups. However, with the group I had, made up of just a few turkeys and a few chickens, after they all got used to each other, more often than not my turkeys would intentionally hang out with the chickens. The chickens slept inside their coop at night, and they didn't come out of their coop in the morning until the automatic chicken coop door opened. On dark, rainy mornings, the automatic chicken coop door didn't open as early as it did on a sunny morning.

When the turkeys flew down from their outdoor roost each morning, they looked around for their chicken friends, and if the automatic coop door wasn't open yet, Eleanor and Prudence would start calling for the chickens. The turkeys stood right outside the door of the chicken coop, yelping their mournful turkey calls until the door opened and they were reunited with their chicken companions.

On stormy days, the chickens would take shelter from the weather and hunker down inside or under their coop, while the turkeys tended to be a bit more tolerant of the nasty conditions and continued to stay out in the open. As they did in the mornings, Eleanor and Prudence would start calling for the chickens when they noticed they were alone in the pasture. Eleanor and Prudence would then go in search of the chickens, and the sad turkey calls would cease only when the turkeys located their chicken friends. I'm not sure whether Eleanor and Prudence thought that the chickens were turkeys or just granted them honorary turkey status, but either way the chickens appeared to be part of the turkey tribe.

CHAPTER 2

What to Expect When Keeping Turkeys

There is so much to enjoy about raising turkeys. In addition to providing a source of meat and eggs, they offer huge entertainment value, and there are so many beautiful heritage varieties to choose from. The day-to-day experience of raising turkeys is very rewarding, and many people form a special bond with their birds.

Turkeys as a Source of Eggs

A BEAUTIFULLY SPECKLED TURKEY EGG is lovely to behold! The shell color varies from light cream to light tan, and there is much variation in the size and pattern of light brown speckles on the eggshells. Turkey eggs can vary in shape from the familiar oval of chicken eggs to more pointed on one end or even almost round.

Turkeys have a shorter laying cycle than chickens because they have not been bred to maximize egg production in the way that chickens have been. Turkeys generally lay eggs for approximately six months of the year. In western Oregon, my turkey hens start laying in late March or early April and continue into early October. The egg-laying cycle is triggered by the lengthening hours of daylight. Turkeys may start laying in late February or early March in other parts of the country, where longer spring days arrive earlier, or when spring weather is especially favorable.

You can expect most turkey hens to lay about 100 eggs in their first year.

Hens typically start laying eggs in the spring following the year in which they were hatched, although there are always exceptions, and I have heard of hens starting to lay in the fall. Just as with chickens, younger turkeys lay more eggs, with production gradually decreasing with age. You can expect most turkey hens to lay about 100 eggs in their first year, or up to 120 eggs for a good layer. Hens that go broody (sitting on eggs to hatch them) will lay fewer eggs in a season, but you can still expect several dozen eggs in the spring before broodiness begins, with another several dozen eggs in the late summer after the broody cycle has

The unique speckle pattern of turkey eggs makes them extra fun to collect!

ended. Some hens have a stronger instinct to be mothers than others. On average you can expect about half of your heritage turkeys to go broody in the spring or summer each year.

Selling Eggs

Regulations for selling eggs to the public vary from state to state. If you plan to sell your eggs, check with your local department of agriculture regarding the rules where you live.

Providing Nesting Places

Before my first turkey hens started laying, I spent a lot of time deciding what sort of nest boxes we should build for them. After they started laying I realized I could have saved myself the trouble because they ignored pretty much everything we constructed. Wild turkeys make their nests on the ground, often under a dense canopy of brush but sometimes in tall, herbaceous vegetation. It is pretty common for domestic turkeys to follow that instinct and ignore the safe nesting places you provide for them. Nesting in the wild is a dangerous proposition, as turkeys are an easy meal for any passing predator. Broody turkeys especially are at high risk of predation. It's much safer for them to nest closer to home, so I recommend giving them a few options to encourage them to lay their eggs where you can keep an eye on them.

Our turkeys and chickens live in a yard surrounded by electric poultry netting. A combination of us mowing and the birds grazing keeps most of the vegetation low. However, we leave a patch in the outskirts of the turkey yard where the grass and weeds grow tall. This is a favorite egg-laying location for the turkey hens. They make a couple of hidden nests that multiple hens use.

It's comical to observe turkeys going through many of the same silly nest box antics seen in chickens. Even with multiple suitable nesting locations, turkeys tend to lay in the same communal nest. Another favorite laying spot for our hens is a stainless-steel dishwasher tub—dubbed "the spaceship"—set near the outskirts of the pasture. I've seen three turkeys sharing the spaceship at the same time in various stages of being broody and laying eggs. Our turkeys also regularly lay in a couple of small coops that were formerly rooster bachelor pads.

The large wooden nest boxes that we custom built to what we considered to be perfectly comfortable turkey dimensions have been ignored,

Some turkeys will squeeze themselves into a chicken nest box to lay their eggs, even if you provide them with a larger nest box.

although a couple of my turkey hens will lay eggs in the much smaller chicken nest boxes. So, before you go to the trouble of building any sort of laying structures, try cultivating an overgrown, secluded spot in your poultry yard and looking for items around your yard or scrap area that could be repurposed into potential nesting sites.

Cooking with Turkey Eggs

Turkey eggs are wonderful for cooking. The higher yolk-to-white ratio lends a rich flavor and gives a little extra rise to baked dishes. The flavor is very similar to that of chicken eggs, but turkey eggs have a higher fat and protein content, making for a hearty breakfast. Scrambled turkey eggs, omelets, and eggs over easy on toast appear regularly on our breakfast table.

When substituting turkey eggs for chicken eggs, dishes that are less fussy, such as quiches or frittatas, are more forgiving, and turkey eggs can be easily substituted for chicken eggs in pancakes with no adjustment to the recipe. For most baked goods, however, you'll need to weigh the turkey eggs on a kitchen scale to make sure you are using the correct equivalent of turkey eggs for the number of chicken eggs called for in a recipe. Having too much or too little liquid can affect the texture and consistency of cakes, quick breads, custards, and so on.

Most recipes assume you will use large chicken eggs. Egg sizes are defined by the US Department of Agriculture and are based on the weight per dozen eggs. The minimum weight of a dozen large chicken eggs is 24 ounces, with each egg required to be a minimum of 2 ounces. The next size up is extra large, with a weight per dozen of 27 ounces and each egg a minimum of 2.25 ounces.

Heritage turkey eggs are on average 50 percent larger than chicken eggs and generally weigh in the range of 3 to 3.5 ounces, with the occasional whoppers being close to 4 ounces. They also have thick shells

Turkey eggs are typically larger than even jumbo chicken eggs.

Comparing Egg Size

Chicken, large	2.0–2.25 oz.
Chicken, extra large	2.25–2.5 oz.
Chicken, jumbo	2.5 oz.
Turkey	3–4 oz.

that can be difficult to crack, so be prepared to give them a good whack! When substituting turkey eggs for chicken eggs, weigh the turkey eggs in the shell to get the equivalent ounces for the number of large chicken eggs called for in the recipe. If you have trouble getting the correct weight equivalent, you can use bantam chicken eggs, the smaller eggs of newly laying chickens, or a partial amount of a turkey egg to get to the correct total weight for the recipe.

Raising Turkeys for Meat

BROAD-BREASTED TOM TURKEYS CAN WEIGH as much as 50 pounds at maturity, while heritage toms typically reach a mature size of 20 to 25 pounds. After a turkey is dressed (plucked and eviscerated and ready for cooking), you can expect it to weigh approximately 70 percent of its live weight. Harvesting turkeys on the homestead requires the same preparation, equipment, and process as harvesting meat chickens. I won't go into detail about butchering and processing turkeys, but I do suggest finding a person who has experience harvesting turkeys to assist you the first time. Expect the entire process of setting up your workstation, doing the deed, and cleaning up afterward to take half the day for a few birds, at least until you become more efficient with the process. If you are planning to raise turkeys to sell to the public, be sure to check with your local department of agriculture regarding the rules in your state.

Heritage turkeys have earned a well-deserved place on the Ark of Taste list by Slow Food USA, but for those accustomed to seeing a gigantic broad-breasted turkey on the table, a heritage turkey may take a little getting used to. A cooked heritage turkey has a leaner appearance, with narrower breasts and longer, more slender drumsticks. That said, heritage turkey breasts still provide a decent amount of white meat.

When we raised our own Thanksgiving turkeys, the dressed birds were in the 12½- to 14-pound range after seven months of growth. Heritage turkeys make a delicious addition to a holiday feast, and everyone who had the pleasure of tasting one of our turkeys appreciated the time, effort, and care that went into providing this special meal.

Turkeys Are Fun to Have Around

I'M SURE YOU'VE REALIZED BY NOW that I love turkeys because they are so engaging, not because they taste good! Despite their main claim to fame, turkeys are so much more than a meal, even a holiday one. No matter what variety you decide upon, your reasons for keeping them, and whether you raise a few or too many to count, I guarantee that you will become enamored with these charismatic birds.

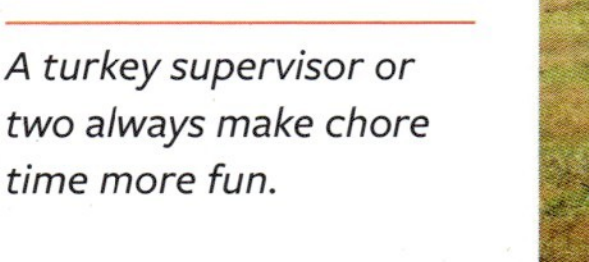

A turkey supervisor or two always make chore time more fun.

Observing their interactions within the flock, with other animals on the homestead, and with their human caretakers is a daily source of entertainment and amusement. I know several non-meat eaters who keep turkeys purely for the joy of having these delightful animals in their lives.

The inquisitive nature of turkeys means that they are always coming to investigate whatever activity is happening near them. Farm chores or improvement projects will almost always draw their attention, and soon you'll have a turkey supervisor watching your every move. Turkeys also show great affection toward the people with whom they have bonded. You really can't ask for better companions than that.

Playing Games

Turkeys are social animals with very curious personalities, and they enjoy playing. Here are two games I play with my friendliest hens.

For the game of "feather," I pick up a feather off the ground and brush it on a hen like I'm tickling her. She grabs at the feather repeatedly, and when she manages to pull it out of my hand, she runs around shaking it vigorously, occasionally dropping it and picking it back up again. It's like she's showing that feather who's boss!

The game of "egg roll" involves putting an egg in front of one of my especially curious hens. She usually begins pushing it around with her beak, trying to roll it under herself. It seems that she wants the egg under her rather than sitting out in the open, and she won't take no for an answer!

PUMPKIN PIE: A TRUE "LAP TURKEY"

One day, watching the feed from the camera we have set up inside our coop, I noticed that one poult was not doing well. When the momma would get up and move to another area, all the poults would follow her except for this one, which was left lying on its back waving its feet in the air frantically trying to right itself. I repeatedly went out to put the poult back under the momma and hoped for the best.

The same thing happened the next day, so I set up the poult in a brooder in the house. I suspected that her problem was the result of not getting enough nutrition during development in the egg. I decided to provide vitamins in her water and give her plenty to eat, in the hope that after a few days she would be able rejoin the flock.

Within an hour of getting the poult settled and feeling like everything was under control, a windstorm knocked out our power! I lit a fire in the woodstove and moved the brooder in front of it. After a couple of hours hanging out there, I was becoming quite attached to this little one. After the power came back on, we settled into a routine. Throughout the day I would check on the poult and encourage her to eat and drink. This was especially important since she didn't have her turkey momma to teach her.

I often heard a loud peeping when I wasn't in the room. When I checked, she was just fine, standing on top of her stuffed animal companion and peeping happily as if to

announce her climbing achievement. In the evenings we would sit on the sofa together, and after a couple of days, I stopped worrying that she wouldn't survive and let myself love her. I named her Pumpkin Pie. After the second night of loud peeping, my husband set up a metronome next to her brooder, which seemed to provide some soothing companionship and quieted her down. That's when I knew that little Pumpkin Pie had worked her way into my husband's heart, too.

After five nights in the house, Pumpkin Pie was much better. I wanted her to be accepted by her family, so I didn't dare keep her inside any longer. When I took her back to the coop, her momma and siblings were busy eating and running around, so I quickly put Pumpkin Pie in with the others and no one seemed the wiser.

I went back inside hoping that Pumpkin Pie would remember me now that she was back with her turkey family. I was thrilled to find that every time I went out to the turkey yard, Pumpkin Pie would come running up to me. I would kneel and lay my hand on the ground, and she would sit in my hand and let me pick her up. The other poults would only spend time with me if I had treats. Pumpkin Pie was content to sit with me for as long as I liked, no strings attached. I finally had the lap turkey I'd hoped for, and I couldn't have been happier.

Pumpkin Pie and I share a special bond to this day.

CHAPTER 3

Turkey Behavior

The conversation really gets interesting when you start talking about the things turkeys do! Turkeys are incredibly charismatic birds, in both appearance and behavior. Whether you decide to raise heritage turkeys or broad-breasted turkeys, and whatever their intended purpose is, they will charm their way into your heart. Their larger-than-life personalities are the reason that turkey math (the urge to keep adding to your flock) is just as real as chicken math.

Though there are some similarities in how turkeys and chickens are raised, as well as in their behaviors, it's important to understand that turkeys are not just big chickens. Turkeys will require a bit more work on your part than a chicken flock will to keep the peace and ensure that everyone is getting the care they need. But I can assure you that the extra effort is well worth it, and they will more than repay you with the amount of joy they bring you.

Turkeys do everything in a big way. They aren't always as easy to care for as chickens are, and it's not as easy to convince them to do what you want them to do. Here's an introduction to what you will observe when you add turkeys to your homestead.

Parts of a Turkey

Although much of their anatomy is similar to that of chickens, features that distinguish turkeys from other types of poultry include their facial structures and the beard of tom turkeys.

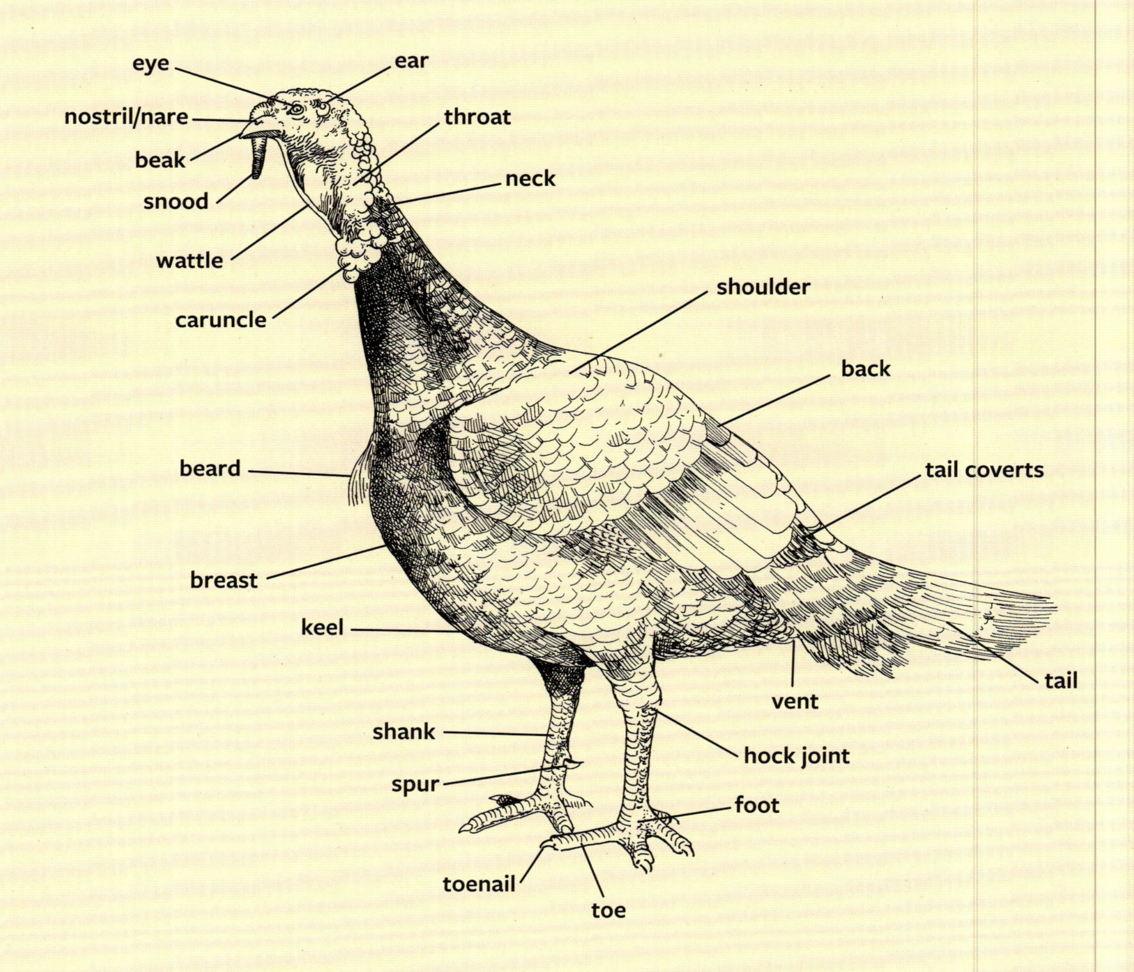

Turkey Anatomy

BEFORE WE TALK ABOUT TURKEY BEHAVIOR, a brief discussion of their anatomy will help you understand why turkeys have all those lumps and bumps and other interesting features.

BEARD. All tom turkeys have a beard, which is a cluster of coarse, bristlelike modified feathers that hang down from the breast. The beard appears as a tom nears maturity and can grow up to 10 inches long. Female turkeys also occasionally have beards, though they are thinner and shorter than the male beards.

CARUNCLES. These fleshy bumps occur on the head, along the neck, and at the base of the throat. Both males and females have caruncles, but those on males are larger and more prominent. The largest caruncles are at the base of the throat and are called major caruncles. The major caruncles on both males and females become bright red when they are excited or angry.

A tom's wiry beard can grow up to 10 inches long.

Both sexes have caruncles. Shown here are the smaller caruncles on a female.

SNOOD. Instead of a comb, turkeys have a fleshy protuberance that originates above the top of their beak. The snood of a tom turkey is one of his defining characteristics, hanging down the side of his beak for several inches. The snood may be retracted when a turkey is relaxed, sleeping, or eating, but it typically hangs down for much of the day. When retracted, the snood is less than an inch in length. The snood becomes bright red when a tom turkey is displaying or feeling aggressive. Female turkeys also have snoods, although theirs are no more than a half an inch in length, they do not extend them, and they tend to be pink.

SPUR. This sharp, pointed protrusion from the shank of the leg in male turkeys is used for fighting.

VENT. Also known as the cloaca, this is the exterior opening for the digestive, urinary, and reproductive tracts and is used to expel feces and lay eggs.

WATTLE. A wattle is a thin flap of loose skin that hangs down below the beak. Both males and females have a single wattle. It is larger and generally a brighter red color in the males.

Left top: *The retracted snood of a tom turkey resembles a tiny unicorn horn.* ***Left bottom:*** *The longer the snood, the more handsome a tom turkey is to females.* ***Center:*** *The snood of an adult hen is considerably smaller than that of a tom.* ***Right:*** *Males use their sharp spurs to fight each other.*

Understanding Facial Colors

The colors of a tom turkey's face and neck change with his mood. A calm tom turkey tends to have paler facial and neck coloration; the face is mostly blue while the back of the head, neck, caruncles, and snood vary from pink to light red. When a tom gets excited during courtship or when he is feeling aggressive around another male, his head, neck, caruncles, and snood turn a bright red. A very light, almost white head also indicates that a male is excited or angry.

The facial colors of a tom turkey can vary from blue to pink to dark red or almost white, depending on his mood.

Tom turkeys put on an impressive display to attract mates and intimidate rival males.

Strutting

THE FIRST TIME YOU SEE A TOM TURKEY STRUTTING, you won't be able to take your eyes off his beautiful display. When a tom turkey struts, he fans out his tail feathers and holds his wings outstretched and slightly away from his body, with the tips of his flight feathers touching the ground. He takes a few steps forward, turning his body and his tail feathers this way and that, showing off his plumage to full advantage. Toms strut as a display of dominance to other turkeys and to humans. If a tom turkey's strut is directed at a rival, the behavior can escalate into more aggressive actions or even full-blown fighting in which the two fly at each other, flogging with their wings and biting their rival's head and neck. The strut is also used as a courtship display and as a precursor to mating.

Some male turkey poults will begin strutting when they are only a couple of weeks old, as they begin to establish the dominance hierarchy with their hatch mates. The sight of a young poult strutting will not only confirm that this youngster is male but also melt your heart.

Strutting behavior is used as a dominance display by some females, too. When you see a turkey hen strutting, it is typically because she's either mad or jealous. A hen may strut if she sees someone else getting treats or special attention, and she may strut and make an angry-sounding call when she's protecting her little ones from a perceived threat, or if she is unreceptive to a young tom turkey courting her.

Wild Instincts

Their wild instincts make turkeys fascinating but occasionally frustrating!

TURKEYS RETAIN MANY OF THEIR WILD INSTINCTS, which is one of my favorite things about them. It's the reason they can be difficult to train to sleep in a coop, the reason you'll often find them roosting where you wish they wouldn't, and the reason you'll feel like they are often playing hard to get. Heritage turkeys that have been hen-raised (rather than hand-raised by a human) often prefer to keep a little separation between themselves and their human caretakers, and they tend to be apprehensive about being touched.

Most turkeys are also very curious, and if you sit down and wait for them to come to you, one or two of your braver turkeys will usually approach you and may even begin pecking curiously at your clothing. Turkeys often fixate on and repeatedly peck at loose-hanging items such as shoelaces and drawstrings or anything shiny such as zippers, snaps, and jewelry.

Turkeys can be trained with treats to come up to you, and the bolder ones may even eat out of your hand. But the moment you try to pet or snuggle them, all but the friendliest ones will quickly move just out of arm's reach. Although their wild instincts may be frustrating on occasion, they also will fascinate you every day and will give you the deepest respect for how self-sufficient they are. It's also what makes it more rewarding when you do manage to earn their trust and you get that first snuggle with a handsome tom or lap time with a sweet turkey lady.

Broad breasted turkeys seem to retain fewer of these wild instincts than do heritage turkeys—perhaps not so much in terms of trying to sleep outside, but more so when it comes to their behavior around people. In talking with fellow turkey owners, the consensus seems to be that broad breasted turkeys tend to be more people-friendly from the start, but if you spend some extra time socializing your heritage turkeys and getting them accustomed to being handled, they can be just as affectionate as broad breasted turkeys.

WILD AT HEART

When we had our first four young tom turkeys, one or two would occasionally jump over their pasture fence. They could usually be found walking back and forth along the outside of the fence, looking for a way back in. This turned out not to be the case with turkey hens. About a month after we acquired our first hens, Eleanor and Prudence, I noticed that Eleanor was not inside the fenced yard when I did my evening rounds. I searched along the edge of the adjacent brush and forest, but she was nowhere to be found.

The next morning, I woke early to search for her. Prudence joined in the search by repeatedly calling for her friend with the distinctive turkey yelp, and I soon heard Eleanor calling back from way down the hill in the blackberries. The lost call of a turkey is a plaintive cry, and it was sad to hear the two friends calling back and forth trying to find each other. I began searching in the blackberry patch by walking along the more open areas the deer used, but from what I could tell Eleanor was at least 25 feet further into the blackberries than I could go.

Eventually I caught sight of her, but whenever I got within a few feet of her, she would head deeper into the blackberries. I gave up on trying to catch her but kept checking on her, and after a few hours, she finally returned to the pasture. Eleanor walked along one side of the fence while Prudence walked next to her on the other side, both calling happily to each other. It took some time, but eventually I chased Eleanor back inside the fence where she belonged.

About a week later, I discovered Prudence was missing. Luckily, it was daylight this time. I could hear her down in the blackberries, flapping noisily and calling. Off I dashed again into the blackberries, where I spotted Prudence perched on a branch.

Prudence was tame enough to eat out of my hand, but she still wasn't tame enough for me to pick her up. I crept close enough to grab her and get her back to the fenced pasture without too much trouble, though there was a great deal of flapping!

Although we had given the turkeys a nice coop and several covered roosting structures, we began to wonder if the ladies were jumping the fence in search of something else. During the day they would take shelter from the rain, but at night they would find a roost out in the open, typically on the highest structure available. We built them a higher roost with multiple perches, and this time we had a winning design that everyone was willing to sleep on.

This structure satisfies our turkeys' wild instinct to roost high off the ground at night.

A 2×4 board placed wide side down makes a perfect turkey roost.

Suspicious Minds

TURKEYS ARE QUITE SUSPICIOUS, and they view anything new with great interest. This is true of new surroundings, new members of the flock, and even the smallest of unfamiliar items such as a garden tool or a pair of gloves. I think this behavior stems from their wild instincts and their strong urge for self-preservation. Anything unusual is viewed as a potential threat, and it is typically announced with a series of alarm calls and a gathering of turkeys circling the newfound object with outstretched necks.

It doesn't matter if the suspicious object is something they've seen before that you simply set down in a different location. If it doesn't belong in that location, the turkeys will have something to say about it! It's quite comical.

Should you see a semicircle of turkeys staring at the ground and making alarm calls, you may find a snake that's hoping to make a getaway before someone gets bold enough to start pecking it. I've found dead snakes in the yard, so I think turkeys will kill them but not eat them.

Unfamiliar objects usually result in a gathering of suspicious turkeys.

Flock Dynamics and Dominance

THERE IS A PECKING ORDER IN TURKEY FLOCKS, just as there is in any other type of poultry flock. Both males and females can be quite adamant about enforcing their position in the flock hierarchy. Compared to chickens, the dominance behavior in turkeys tends to be a few notches higher on the sassiness scale; the word *relentless* comes to mind. This is a good reason to give your turkeys as much space as possible; the ones that are lower in the pecking order may appreciate having areas where they can spend time away from the more dominant flock members. Having plenty of space is also important in a mixed flock of poultry, especially if you have observed bullying behavior by your turkeys.

How Many Tom Turkeys Can Coexist?

By the time young male turkeys, known as jakes, reach four or five months of age, they will begin displaying and challenging each other to establish the flock hierarchy. Displays become more intense by the time they are about six months old. Male turkeys can fight viciously, and it can be very hard to watch as they struggle to determine which will be the dominant male. Fighting toms will fly at each other feet first, attacking with their toenails and spurs in much the same way that roosters fight. Toms will also use their wings to flog each other, and they pack quite a punch with all that weight behind them. Should you ever be on the receiving end of a wing slap to the face, you'll do your best not to let it happen again.

Allowing toms to establish their rank is important for flock dynamics.

Tom turkeys will also bite and pull at a rival's face and neck, refusing to let go for what can seem like an eternity. Sometimes one tom will even manage to put his mouth around another tom's head, and they can become interlocked at the risk of suffocating them both. Watching these interactions is not for the faint of heart, but allowing them to play out is important for the flock dynamics.

If it looks like things are getting out of hand, however, you can try to break up the fight, although it's not easy once the toms are seriously engaged. Spraying the fighting toms with a hose or throwing a bucket of water at them has no effect. You may need to physically separate them and put them into separate enclosures or on opposite sides of a fence for a time-out, although even that won't entirely stop the fighting if they can

Multiple toms may be able to coexist just fine, depending on their temperaments.

still see each other. They will need to be visually isolated from each other to truly settle down.

Should the fighting escalate to this point with no apparent establishment of the hierarchy coming anytime soon, you may decide to rehome the extra males in your flock or send them to freezer camp if that's one of your reasons for getting turkeys in the first place. Depending upon the temperaments of your males and the setup of your turkey yard, it may or may not be possible to keep more than one tom turkey in your flock.

For many years, I kept only one tom turkey year-round in my flock because I couldn't stand the fighting. When that tom got to be six years of age, I found that he tolerated having one of his sons remain in the flock. His son didn't challenge his authority for several years, but eventually, his son and a younger tom joined forces to defeat him. It is the natural way of things, although that doesn't make it any easier.

Some people can keep multiple toms in their flock, and they can live together relatively peacefully. Other people build separate pens to keep the peace, with each tom having his own area and his own group of ladies. Everyone's situation is different, so be prepared to make modifications to your setup as your tom turkeys mature and you see how the flock dynamics change over time.

Tom-to-Hen Ratio

I recommend keeping three or more turkey hens for every tom turkey, especially if you plan to keep them year-round and for breeding. Heritage turkey hens tend to go broody often, so while your hens are broody or raising young, it's a good idea to have some other turkey hens around to occupy the attentions of your tom turkey and keep him from becoming a nuisance to the other poultry in his flock or to his human caretakers.

You Are the Boss

FLOCK DYNAMICS ARE NOT ONLY ABOUT THE HIERARCHY among turkeys and their feathered pasturemates. They are also about turkeys' relationship with their human caretakers, because their people are part of the flock, too. Tom turkeys have quite an imposing presence, and they can be more than a little intimidating. It's important to let your tom turkey know from an early age that you are the boss in the relationship. Use your body language to maintain your personal space, and always let tom turkeys know when they are overstepping the boundary between acceptable and unacceptable behavior.

Tom turkeys can be intimidating, so establish your dominance early.

Here is an example. When I sit down in the turkey yard to enjoy a few minutes of watching my turkeys (I call it "Turkey TV"), my dominant tom turkey is quick to come over and begin displaying for me. He's fun to watch, and he will get progressively closer and closer to me. If I have my legs outstretched in such a way that he could mount them (as he would do with a hen), he will often attempt to do so. While this can be either entertaining or mortifying, depending on your view of acceptable behavior, it's not a good idea to allow it. You are not your turkey's mate, and he should not view you as being subservient to him.

When my tom starts exhibiting this behavior, I gently push him away, change the position of my legs, or stand up. Depending upon how persistent he is being, I may force him to move away by walking slowly but firmly toward him, thereby retaining my dominance in our relationship. Remember to be consistent, as turkeys like to test boundaries.

Tom Turkeys Behaving Badly

One of the most common laments I hear from new turkey owners is "My tom turkey used to be so nice, and now he is attacking me/my spouse/my child. Why is he doing this, and what can I do to stop him?" I wish I had a better answer to this question than "Unfortunately, not much." Tom turkeys can become possessive of their primary caretaker, and they may get jealous and act out toward other members of their main caretaker's family.

This happened with my husband and my dominant tom turkey. My husband occasionally spent time with this turkey as he was growing up and hand-fed him treats until he was several months old. But one day the tom mysteriously began attacking my husband, and we haven't been able to change his behavior, despite trying the techniques mentioned in the following paragraphs.

Thankfully in our case, the attacks are more of an annoyance than a full-on flogging. When my husband goes into the turkey yard, he knows to keep an eye on my tom and to block or avoid any attempted attacks. Sometimes my tom turkey is just in the mood for trash talking and won't make a serious effort to attack. If he is being more troublesome, I will escort him to the other side of the turkey yard fence or run interference and distract him so that my husband can do what he needs to do without any sneak attacks from behind. It's annoying for my husband, but it's manageable, and it always gives us a funny story to tell. (Well, at least I think it's a funny story!)

This tom turkey's aggressive body language is clear in a standoff with his caretaker, whom he is treating as a rival.

This can be a frustrating situation, however, so here are a few pieces of advice. First, always wear proper clothing in the turkey yard, which means no shorts or open-toed shoes if you have a tom that is prone to attacking. Second, if this aggression is a shift in your tom's behavior, try to figure out if something in his environment has changed that could

account for the difference. Is the tom's favorite hen broody or off raising poults? This may contribute to his irritability, and maybe just giving him some extra space during this time will help. Did you get a new brightly colored pair of boots or a new article of clothing that may be setting off your tom turkey? Maybe a wardrobe change is all that is needed.

Third, you can try to reassert your dominance by using the same types of techniques that would be used for an aggressive rooster. These include handling him when he exhibits aggressive behavior. Don't fight him, as that will only escalate the situation. Calmly stand your ground and either hold him down or pick him up in your arms for several minutes to subdue him. He may calm down, or he may not—once a tom has the idea that he's dominant in the relationship, there's not much that is going to change his mind.

As a last resort, when you go into the turkey yard, carry something long, such as a broom or rake, that you can use to shield yourself from an aggressive tom if needed. I'm not saying you need to use it to fight off an attack, but just having something to hold between you and the offending tom is often enough of a deterrent. If young children are the object of your tom's aggression, then it's time to think seriously about either keeping your tom securely fenced or rehoming him.

Gently but firmly holding an aggressive tom turkey may help you assert your dominance.

Turkey Talk

THE MOST RECOGNIZABLE SOUND made by turkeys is the gobble. Only tom turkeys gobble, but both males and females are very communicative, and they make a variety of vocalizations, which is another reason they are so fascinating. I've read that wild turkeys make 28 distinct sounds, and although I can't distinguish the subtleties of that many sounds in my turkey yard, here are the more common ones that you'll be able to learn easily.

Gobble

The gobble is the most frequent sound made by tom turkeys, who gobble for several reasons. It is used as an alarm call as well as to express dominance. A tom will often gobble when a potential threat is spotted, either in the air or on the ground. My tom turkeys may gobble to announce visitors to the farm or upon hearing a nearby loud noise, whether it be a delivery driver or the garbage truck down the road. Interestingly, my toms rarely gobble when they see our car or truck drive up, but they often gobble when an unfamiliar vehicle approaches.

Turkeys employ a number of distinct vocalizations and are quite communicative.

Toms usually gobble when people come into view, which is their way of expressing either their dominance or excitement. When my boys see me walking out with treats in my hand, they start gobbling in unison, which is fun to hear. You get the idea—there's usually a whole lotta gobbling going on when you have turkeys.

Spit and Drum

Another common sound made by tom turkeys is known as the spit and drum. This low, two-part sound can be difficult for humans to hear and describe. Visitors often ask me about it when one of my tom turkeys is strutting up close. When tom turkeys are displaying, they make a short "spit" sound, followed by a soft, very low-pitched drumming that sounds like a quiet rumble. The combination sounds something like *Pffit, druuummm.*

Purrs

Both male and female turkeys make a couple of types of purring sounds. A soft purr indicates contentment. Sometimes I'll hear one of my my tom turkeys purring happily with his head buried in the feed pan. I'll often hear my turkey hens purring when they are dust bathing. As well as indicating contentment, purring also seems to be used to communicate

A wild turkey in mid-gobble

with the rest of the flock when they are not in eyesight of each other to let everyone else know "I'm fine and I'm over here." Turkey mommas also make a purring sound to gather up their poults at bedtime and encourage them to come back to the nest for the night.

A second type of purr, known as the "fighting purr," is a higher-pitched, more insistent sound than the purr of contentment. As the name suggests, turkeys emit a fighting purr when they are actively engaged in challenging a rival, but ironically it is an adorable sound that doesn't seem at all what you would expect from a mad turkey. When a tom turkey gets really worked up, the fighting purr is often interspersed with rapid gobbles. There's no mistaking the sound of an angry tom turkey. No matter where I am on the farm, I always know in an instant when my husband is visiting the turkey yard because my dominant tom will immediately start gobbling loudly at him, mixing in a lot of fighting purrs for good measure.

Turkey hens make a version of the fighting purr, and it's equally adorable, although it means that someone is in a bad mood! My turkey hens often get riled up when the toms are performing a courtship display or are mating with one of the other hens. It almost seems as if the hens that are being ignored by the toms are jealous of the other hens getting all the attention from the boys. A big hen chase often ensues, with many fighting purrs heard from the dominant hens. A momma turkey will also make a fighting purr when another turkey comes too close to her poults, meaning "You better back off or there's gonna be trouble!"

Hens have a variety of calls, especially when they are keeping track of their poults.

Miscellaneous Calls

Turkeys, in particular adolescents and hens, make a variety of other calls. In the context of wild turkeys, these calls are described as cackles, clucks, cutts, kee kee, putts, tree calls, and yelps. Both heritage and broad breasted turkeys make a variety of these calls, although some of them are similar and may be hard to distinguish from each other. The yelp is one of the most common sounds; to me it sounds like a series of short barks. This sound indicates that a turkey is worked up about something.

Turkeys yelp when they are in distress, such as when a bad storm is blowing in or if someone has become separated from the flock. It is used as a homing or lost call, with both the separated individual and the rest of the flock calling back and forth until they are reunited. Sometimes I'll hear a hen yelping as she wanders the turkey yard looking for a suitable place to lay her egg and not being satisfied with the options.

The alarm call made by hens is a high-pitched single-note call described as the "putt" sound, but I personally think it sounds more like "pip." This loud call is repeated in quick succession. Hens make this sound when a potential threat is spotted, such as when an aerial predator flies overhead or a hawk lands in a nearby tree. Often several hens will chime in and begin alert-calling in unison to warn everyone about a potential danger. The alarm call is a distinctive sound, and when I hear it, I usually run outside or look out the window to see if there's something urgent I should attend to.

In addition to the yelp and alarm calls, turkey hens use a variety of other calls throughout the day. They have a few types of short, sharp calls that may be uttered as a single note or rapidly repeated a few times. These calls are described as cutts and clucks (although a turkey's cluck doesn't sound anything like a chicken cluck), and these sounds are used to get the attention of another bird or when a turkey is excited. While I'm sure there are subtle differences in the inflection of the call and what that means to others in the flock, for the most part when you hear these calls there's nothing to worry about; it just means that a chatty turkey is voicing her opinion about something.

When turkeys are on the roost in the morning, they will often talk among themselves. They will also call as they fly down from the roost. The sounds they make in the morning or when flying down from the roost are like a yelp or putt sound, but with less urgency. These sounds are known as tree calls or fly-down cackles. At first when you hear these calls in the morning you may think they are alarm calls, but they can be distinguished from alarm calls because the fly-down call is usually just a single call or two, whereas an alarm call tends to be repeated for a longer time.

Courtship and Mating

THE COURTSHIP DISPLAY OF A TOM TURKEY is a beautiful sight to behold. He will strut while taking a few steps forward and then turn from side to side to side, rivaling a flamenco dancer in the impressiveness of his dance moves. Upon seeing this magnificent display, a turkey hen will usually squat down on the ground and sit perfectly still, waiting for the tom to dance his way over to her and proceed to do the deed. The mating act itself can take some time, and you may find yourself feeling a bit sorry for the poor hen patiently waiting for the big lug to finish and get off her.

A group of wild turkey toms strut their stuff to impress hens.

The mating ritual of turkeys can be a protracted affair, sometimes taking several minutes.

Tom turkeys have the habit of treading on a hen's back for what seems like an eternity.

Those unfamiliar with turkey mating may mistake the treading behavior for a type of attack. Rest assured that treading is perfectly normal. It can be particularly amusing to watch the younger toms, as they often have difficulty figuring out which way to point their business end and they will tread on the hen's back for the longest time while facing in the wrong direction. You may find yourself thinking there is no way you'll ever get fertile eggs from this unskilled performance. But toms are quite persistent, and eventually they figure out how to get the job done.

Generally, watching the mating ritual of turkeys is less stressful than observing an overly amorous rooster. Tom turkeys don't chase down the hens as roosters often do, and typically the act appears to be mutually desired by both turkeys. A couple of tom turkey mating behaviors, however, can be a bit troublesome to watch. After the mating act has been completed, toms have a strange habit of continuing to sit or stand on top of the hen for no apparent reason other than to continue squishing the poor girl! I haven't been able to figure out the purpose of this behavior other than perhaps it being some sort of display of dominance to anyone that happens to be watching.

There have been occasions when I've seen a tom turkey grab the head of the hen during mating and pull on her skin, which is obviously painful for the hen and sometimes she will begin calling out. I have sometimes interrupted the mating when the hen appears to be in distress. But oddly enough, even though you would think the hen would be anxious to get away from the tom, she will often squat back down and the whole mating act will begin again. I find it's better to just let things run their course rather than interfering, which tends to make the entire mating process take longer in the end.

CHARISMATIC SUBJECTS

Turkeys are fun to photograph, and soon you will find yourself with a camera full of images. Especially curious turkeys will come right up to you to investigate when you point your camera or phone at them, and you'll have lots of hilarious close-ups and bloopers as they peck at it trying to figure out what it is.

Tom turkeys in particular love to show off and ham it up for the camera, and why shouldn't they? The camera loves them! If you are a poultry lover like me, you may find yourself documenting seasonal events on the farm with fun turkey photos and even trying to take family pictures with them for your holiday cards. Here are a couple of my favorites.

Christmas on the farm provides an opportunity for dress-up, if you have a turkey that will tolerate clothing long enough for you to get the shot.

Not all turkeys like to pose for the camera, but the ones that do are usually real characters!

CHAPTER 4

Shelter and Space Requirements

Turkeys have a huge independent streak, and they can be notoriously stubborn when it comes to housing them at night. You may build the most fabulous accommodations, only to find out that your turkeys didn't ask for it and they don't want anything to do with it! Turkeys instinctively want to sleep up high with a good view of their surroundings. They often prefer to sleep out in the open rather than in enclosed spaces. Trees, fences, and rooftops are all fair game when it comes to favorite roosting spots.

Don't be surprised and don't despair if a few of your turkeys prefer to sleep outside even on cold and stormy nights. Turkeys are amazingly hardy, capable of withstanding very unpleasant weather conditions thanks to their water-resistant outer feathers and many layers of downy under feathers. But we all feel better knowing those under our care are tucked in safely for the night, so let's delve into a turkey's needs and wants and try to set you up for success in providing shelter for your turkeys.

An open-sided barn-type structure is a good option for sheltering your turkeys.

Do Turkeys Need a Coop?

THE ANSWER TO THIS QUESTION depends on several factors. Are you planning to raise broad breasted turkeys for the Thanksgiving table, or do you want to keep heritage turkeys year-round? Will your turkeys free-range, or will they be fenced in a secure yard? If you don't intend to keep turkeys through the winter or if they have a secure yard to sleep in, they may not need a coop. Adult birds can live comfortably in most climates if they can get away from the worst of the winter weather.

The answer also depends on whether you are planning to raise poults or get turkeys that are a bit older. If you plan to raise your turkeys from poults, then the answer is a resounding yes, at least until they are ready to live outdoors. However, if you are adding turkeys that are already a few months old to your flock, you may not be able to train them to sleep in a coop.

Broad breasted turkeys tend to accept enclosed life more readily than heritage turkeys, although with both varieties, it is important to begin training them to go into their shelter at night immediately after they move to adult accommodations. It's critical to be diligent about it—even one night spent outside in the great outdoors will likely flip a switch in their brains, and from then on it will be a challenge to convince them to sleep indoors.

Enclosed Housing

I INTENTIONALLY USE THE WORDS "housing" and "shelter" instead of "coop" to emphasize the importance of not thinking of shelter for turkeys as a coop, which I equate with a small structure that turkeys are less inclined to use. If you think big when it comes to housing your turkeys, they just might decide to sleep inside. It helps to be flexible and understand that turkeys will continually challenge your notion of what you think is best for them. You may need to adjust and rework the shelters you've provided for your turkeys once you begin to understand their behaviors. They like to roost high off the ground in a spot where they have a good view and don't feel caged in.

In an enclosed shelter, turkeys typically need much more space per bird than chickens. You may need to provide 15 to 20 square feet per bird. It's possible to get by with less space with flocks that are more bonded or better behaved. For the best chance of success when training your turkeys to sleep indoors, provide a shelter with a high roof and a spacious interior—more like a barn than a traditional coop.

In fact, an existing barn or a large storage shed may be the perfect place to lock your turkeys in safely at night. When repurposing an existing building, consider replacing a wall or a section of a wall with hardware cloth to give it a more open appearance. If you build housing from scratch, consider incorporating one or more screened sides to give it the open-air feeling and line of sight that turkeys crave. Plentiful ventilation is also important to prevent moisture from becoming trapped inside, as well as to prevent the structure from becoming too hot in the summer.

Large adult turkeys are less agile than chickens, so place doors near the ground to provide easy access. If your design incorporates a ramp, build a solid one, and not with rungs that may be difficult for big feet to navigate. Cover the ramp with a nonslip material such as roll roofing. But no matter how much thought and care you put into the design of your turkey shelter, just know that there will undoubtedly be a few birds that prefer to sleep on top of it instead of in it.

TURKEY COOP FAIL

I did a lot of research and felt confident that my first attempt at a turkey coop would be perfect. Yet by the time my first four turkeys reached about three months of age, they had unanimously decided they weren't going to sleep in it any longer. This was when I first learned that turkeys are incredibly independent and almost always do what they want!

I designed my first turkey coop as a half-enclosed coop and a half-screened sun porch. Each half was 50 square feet. My intent was to house three adult turkeys in the coop, as I had planned on keeping a breeding trio consisting of one tom and two hens. The enclosed portion of the coop would provide warmer shelter during the winter weather, while the sun porch would provide open-air accommodations for the summer, giving them the best of both worlds—or so I thought.

I ended up with all males in my first group of four poults, and the building just wasn't large enough for those big-boy personalities. They began sleeping on top of the coop and making a mess of the roof, so we began experimenting with various heights and styles of outdoor roosts and simple shelters until we found a combination that worked for us.

The structures in our turkey yard have evolved from a coop to a covered roost to an open-sided barn.

Simple Turkey Shelters

IF YOUR TURKEYS ARE AS DETERMINED TO SLEEP OUTSIDE at night as mine are, a simple nighttime roosting structure that worked well for my turkeys may be a good fit for your turkeys, too. The structure has several roosting bars about six feet off the ground and a corrugated metal roof. It doesn't have enclosed sides, but it does provide a roof over their heads to keep them dry if they choose to use it. Of course, there are always one or two that prefer to sleep on top of it in all but the worst of weather, but if it's raining or snowing at bedtime, almost everyone chooses to sleep under the roof.

ELEMENTS OF A SUCCESSFUL SHELTER

- High roof
- Unobstructed view of surroundings
- Good ventilation
- Multiple roosts at varying heights

This simple structure with open sides and a roof over a choice of roosts provides adequate shelter for nighttime roosting in all but the worst weather.

Daytime Shelters

As well as providing nighttime shelter, it's important to provide turkeys with options for taking shelter from inclement weather during the day. Turkeys spend much of their time foraging and exploring, but on stormy winter days they will seek shelter. They may use their nighttime shelter, but turkeys are independent and tend to want to spread out, so they may not all want to gather under the same shelter during the day. A couple of simple shelters that provide a place for less dominant birds to retreat from more dominant or aggressive members of the flock can help keep the peace.

If you plan to let your turkey hens raise up the next generation, you will want to provide a variety of small, simple structures that they can use to find some privacy from the rest of the flock. Don't position shelters too close to perimeter fencing or the turkeys may use them as a launching point to fly over the fence.

Daytime shelters don't need to be fancy to protect from the weather while still meeting a turkey's desire for personal space. A simple design with a roof and two or three sides that provide protection from wind, rain, and snow works fine. A metal carport or RV canopy makes an easy-to-install daytime shelter, or you can fashion rustic lean-tos and open-air shelters from scrap wood or pallets.

A simple daytime shelter can be built with pallets and scrap lumber.

Temporary Shelters

You may need a temporary quarantine shelter if you are adding teenage or adult turkeys to your flock. It is important to quarantine any new additions to ensure they are healthy before introducing them to the rest of your flock. Another benefit of keeping new turkeys contained in an enclosed shelter for a few weeks is that it helps them learn where their new home is, so they don't try to fly away immediately. It also gives you a space to spend time with them and start gaining their trust.

When we brought home two young turkey hens for my tom turkey, we built a temporary containment structure using long pieces of ¾-inch PVC electrical conduit bent into tall arches and secured to the ground on rebar pegs, with heavy-duty knotted netting over the top. This allowed me to have some bonding time with the girls inside their shelter while they adjusted to their new home, and when we opened the enclosure after a couple of weeks, they stayed put.

My positive experience introducing new hens using the temporary enclosure method is in direct contrast to the experience of a nice woman who bought an adult turkey hen from me as a mate for her solitary tom turkey. She kept the hen in a coop for a day, and when she let the hen out the next day, her tom turkey liked his new lady friend very much. Unfortunately, the feeling was not mutual, and the hen promptly flew over the fence, ran down to the neighbor's property, and took her new owners on a 1½-hour wild goose—I mean, turkey—chase! So do yourself a favor and keep those new additions contained for a few weeks so they don't immediately fly the coop.

Roosts

THERE ARE SOME IMPORTANT DIFFERENCES between design requirements for turkey shelters and chicken coops. Turkeys' roosting bars should be wider and must be placed farther from the wall than those for chickens. Roosts can be made from 2×4 boards positioned with the wide side up. Branches can make good roosts, but make sure that the diameter is easy for turkeys to grip, and leave the bark on to provide extra traction. Avoid smooth or rounded materials, such as dowels. Those big turkey feet can be clumsy, so your birds will appreciate something they can easily grip.

Most turkeys like to roost high, and after first flying to the lowest roost available they will typically leapfrog up to progressively

A variety of roost heights allows everyone in the flock to find their place while respecting the pecking order.

This roost is made from a 2×4 board, which is a good size for big turkey feet to hang on to.

higher roosts. In most situations, I recommend a lower roost that is three to four feet off the ground; the highest roosts should be six or seven feet off the ground. Exceptions for broad-breasted turkeys are discussed below.

Every chicken keeper knows that bedtime in a chicken coop can be chaotic as flock members jostle for their preferred locations on the roosting bar. The dominance hierarchy is on full display. Turkeys take those nighttime roosting shenanigans to an entirely new level (literally!). Turkeys with bossy personalities may not be content with a comfortable two or even three feet of roosting space. They may decide they want the entire roosting bar all to themselves and will repeatedly peck anyone else with the nerve to try to roost next to them, until the others abandon the roost entirely.

The recommended amount of linear roosting space is 15 inches per bird, but I prefer to provide multiple heights and locations with much more total length of roosting space than is strictly necessary. Having a choice of multiple roosting bars and heights gives the less dominant turkeys places to roost out of pecking reach of the more dominant turkeys. Some people recommend having roosts all at the same height to prevent dominance fighting, but in my experience if all the roosts are at the same height, the turkeys will just go around and around chasing and pecking each other off the roosts. You may need to modify your initial setup as time passes and you see how your turkeys behave.

Roosting Needs for Broad-Breasted Turkeys

Due to their larger size at maturity, broad-breasted turkeys have somewhat different roosting needs than heritage turkeys. Most heritage turkeys prefer to sleep up high throughout their lifetime (the exception being older toms). When they are younger, broad-breasted turkeys will sleep up high as well, but for adult birds, it's a good idea to provide lower roosts, no more than 18 inches off the ground, to prevent injury to their legs or feet when they jump down. Broad-breasted turkeys become less able to roost on a roosting bar as they mature. As adults they will likely choose to sleep on the ground or on the floor of their shelter, or they may appreciate something low and easy to roost on such as a straw bale.

Turkeys don't damage grass the way chickens do.

How Much Outdoor Space?

AS WITH ANY TYPE OF POULTRY, it's best to provide as much space as you can. This is especially true for turkeys, with their big personalities and their propensity for exploring. Having plenty of space to roam and a variety of roosting structures and activity areas helps prevent boredom, which in turn helps keep the peace. One of the nice things about turkeys is that they are not as destructive to grass and vegetation as chickens are. I have my turkeys in the yard in front of the house, and while it would be a disaster zone if it was occupied by a flock of chickens, that's not the case with turkeys. They will make a few dust baths here and there, but in general turkeys won't dig up every square inch of grass in search of delicious insects (or just for the heck of it!) as chickens are so fond of doing.

Our fenced turkey yard is approximately 100 feet by 125 feet, or almost a third of an acre, and our flock ranges from 8 adults up to 20 adults and juveniles depending on the season. This stocking rate works well with our yard setup and the pecking order. It's important to keep in mind that what works for one person may not work for everyone. If you are planning to keep multiple tom turkeys year-round, your space requirement will depend greatly on their personalities and how well they tolerate each other. It's possible that you may need to set up separate enclosures or pens and alternate the toms' access to the outdoors if they are prone to fighting.

Protecting Free Rangers from Predators

MANY PEOPLE CHOOSE to let their turkeys free-range during the day. Turkeys are excellent foragers, and they can find much of their daily intake on their own, although they will still need to be provided with a well-balanced feed to make certain they are meeting their nutritional requirements. Turkeys love to explore their surroundings, and it's fun to watch and listen to them as they enthusiastically go off on an adventure. The risks of the free-range lifestyle can be high, however, especially for young poults, mommas with little ones, and broody hens. Predators include coyotes, foxes, bobcats, raccoons, hawks, owls, eagles, and even your neighbor's dog. In addition, snakes, opossums, skunks, weasels, and rodents will raid nests.

It's entirely possible that your turkeys will not appreciate your efforts to keep them safely housed.

For me, the trade-off between letting my birds free-range and knowing that predators will take their toll is not worth it. I keep my turkeys in a fenced yard surrounded by 4-foot-high electric poultry netting. Although turkeys can easily fly over a 4-foot fence, mine rarely do so. The key is to train them to know where their home is right from the start and provide them with everything they need inside the fence. This will give you the highest likelihood that they will stay put. The size of the yard and our stocking rate ensure that our turkeys always have plenty of green grass and lots of opportunities for foraging, and this setup gives them everything they would have if they free-ranged without any of the risk.

It's entirely possible that your turkeys will not appreciate your efforts to keep them safely housed at night. If your turkeys are devoted outdoor sleepers, then you'll need to take measures to keep them safe from nocturnal predators. Even adult birds high up in the trees can easily fall victim to raccoons. It's always better to be safe than sorry, and electric fencing is an excellent way to keep outdoor roosting birds safe at night. A livestock guardian dog is another option.

Electric Fencing

There are two main options for electric fencing. One option is to build a permanent field fence and install a couple strands of electrified or "hot" wire, one along the top of the fence and a second strand on the outside of the fence approximately six inches off the ground. An easier way to add electric fencing is to use portable electric poultry netting. We use

Electric poultry netting keeps our flock safe from predators.

4-foot-high netting with sturdy step-in posts that are easy to move for mowing or rotating the pasture. Many farm supply companies sell electric poultry netting, at varying price points and levels of durability.

Electric fences can be powered by a solar charger, a deep-cycle battery, or an outdoor-rated electrical cord if your turkey yard is not too far from an exterior outlet. Be sure to occasionally check the charge of your electric fence to be certain that it is putting out enough voltage to deter large predators. If it's not maintaining a charge, you may need to trim the vegetation along the bottom of the fence to stop it from grounding out.

When using portable poultry netting, you must periodically check that the fence posts are still firmly set in the ground. Tunneling rodents and wet or snowy weather conditions can cause the posts to become loose, so be sure to walk the fence line occasionally to check that the posts are secure and the fence is tight.

Free Bird

An electric fence works great to keep predators out, but it won't contain turkeys 100 percent of the time. Given ample space and plenty of food and pasture, turkeys tend to stay inside a fenced area most of the time. However, if turkey hens or young birds become excited during a squabble over dominance or while trying to avoid an overly zealous tom turkey, one or more birds may fly over the fence. The escapee will usually circle the fence repeatedly, trying to get back in, seemingly incapable of flying back over the fence that it flew over only moments ago. On these occasions, I'm glad that I work from home most of the time. Thankfully, whenever one flies over the fence, it is often due to a commotion in the flock, and loud vocalizations usually alert me so I can go outside and shoo the errant bird back inside.

Your last turkey chore of the evening should be a quick check of the flock a little before dusk to be certain that no one is stuck outside the fence at bedtime or roosting somewhere potentially unsafe. Turkeys that are walking the fence line trying to get back in can be shooed back inside by first opening the fence, then circling behind them and walking toward them with your arms outstretched. If you approach slowly, they will usually walk away from you and toward the opening in the fence.

Birds that are already roosting high up are a bit trickier to dislodge. Try to gently nudge them with a long pole or tap on their feet or legs until they get tired of being prodded, at which point they will often fly down, and you can chase them back home.

This escapee seems to be heading off for an adventure outside the fence.

Clipping Their Wings

Young turkeys are excellent flyers, and if they are confined in small areas by fences that are only a few feet feet high, they will begin flying over them regularly by the time they are several weeks old. After you've chased them back inside the fence enough times, you may decide to clip their wings. Clipping the primary flight feathers on one wing should create an imbalance and reduce their ability to fly. Feathers will need to be trimmed again after they molt and grow new wing feathers.

It is not a good idea to clip the wings of turkeys that will free-range, since they need to be able to fly to roost and to escape predators.

I don't clip wings, but here is a basic outline of the steps.

It's helpful, but not necessary, to have two people for this procedure. One person holds the turkey and spreads out the wing for the second person to cut the feathers. Use a sharp pair of scissors with long, sturdy blades.

Cut the flight feathers only—they are the longer feathers on the outer edge of the wing—to approximately half their length. Point the scissors away from the bird's body when you cut.

Be careful not to clip any developing feathers, called pin feathers or blood feathers. The shaft of a pin feather is darker due to the presence of blood and shorter in length than that of developed feathers. If you accidentally cut pin feathers, they will bleed, whereas in fully developed feathers the blood vessels have receded and the feather shaft becomes hollow.

It's good practice to always clip the feathers on the same wing (right or left) when trimming feathers on more than one bird. That way you can avoid accidentally clipping feathers on both wings, which may not be as effective in preventing flight as clipping just one wing.

The proper technique for clipping the primary wing feathers is to remove up to half the length of the longer outer flight feathers on one wing.

CHAPTER 5

Nutrition and Care

Turkeys are excellent foragers, and they are well suited to being raised on pasture, where they can find much of their own food from grasses, weeds, seeds, and insects. On days when the weather is decent and green vegetation is available, it's not uncommon for turkeys to spend much of their day foraging.

Yet no matter how adept turkeys are at finding their own food, you should provide them with a nutritionally complete poultry feed to ensure their dietary needs are fully met. Free ranging alone will not provide them with all the vitamins and minerals they need for proper growth and development. If a turkey's diet is lacking in certain vitamins or minerals, especially within the first few months of development, serious health problems can occur.

Nutrition

THE NUTRITIONAL REQUIREMENTS of turkeys vary at different developmental stages, and different feed formulations must be provided to ensure proper development at each stage. Poults grow much more rapidly than chicks, so they need more protein. (See Chapter 6 for more about feeding poults.)

Proper nutrition is also important for laying hens, especially if you are planning to hatch eggs. For the month or so prior to the start of laying season (which generally coincides with the onset of spring), provide nutrient-dense feed and add vitamins in the water. If your hens have been on a maintenance type of feed through the winter, switch them back over to layer feed and make sure they have access to calcium so their eggshells will be nice and strong. You can think of this as prenatal care for the momma and the next generation.

Though turkeys can forage for a lot of their own food, they still need to be provided with a nutritionally balanced feed.

POULT STARTER FEED. Poults need feed with 30 percent protein, compared to 18 to 20 percent for chicks. Failing to provide a high-protein starter immediately after hatch can result in mortality of poults. If you can't find a feed specifically labeled as turkey starter, you can feed game bird starter, which is essentially the same thing. Provide starter feed until the the poults are six to eight weeks old.

DEVELOPER/GROWER FEED. An appropriate developer/grower feed for turkeys is approximately 20 percent protein. Again, you can substitute a feed designed for game birds, which may be called game bird grower or flight conditioner. The important thing is to provide feed with the appropriate percentage of protein content for each stage of development. Do not provide a feed with calcium until your young turkeys are close to adulthood; too much dietary calcium at a young age can cause kidney damage. Feed the grower ration until the birds are four months old.

Layer feed in whole-grain form

Grower feed in pellet form

Starter feed in crumble form

LAYER/MAINTENANCE FEED. From four months on, hens (and toms, if they are in the flock) can be fed a poultry layer feed with 16 percent protein and higher calcium content in preparation for the breeding and laying season. In flocks where toms are kept separate from females, adult toms can switch to a maintenance feed with a lower protein and calcium content. For the sake of simplicity in flocks with developing birds as well as adults of both sexes, some people continue to feed their turkeys a grower feed as adults. In this situation, a free-choice source of calcium should be provided for laying hens.

Feed Efficiency

This may come as no surprise, but turkeys eat a lot! When allowed to free-range or given access to a productive pasture, turkeys can forage for approximately 50 percent of their diet from the spring through the fall. If your objective is to raise turkeys for the table, it is important to provide high-quality feed to supplement their foraging to get them up to weight in a reasonable period.

I always have feed available for my turkeys—this is referred to as free-choice feeding or free feeding. I have a large hanging feeder and two feed pans that I keep filled with whole-grain layer feed. I top up each feeder every morning with a large scoop of feed. In the afternoon, I throw out several cups of black oil sunflower seeds. There is always food left in the feeders at the end of the day as my flock typically forages a lot.

Turkeys will alternate between eating at the feeders and foraging on their own throughout the day. On nice days when they have spent much

Heavy-duty plastic pans make good feeders. They are easy to clean and can be spread around to minimize squabbles over food.

of their day foraging, everyone usually takes a turn at the feeders in the afternoon to fill their bellies for the night. Rather than using feeders, some people choose to scatter feed on the ground to provide more areas for feeding, which can reduce competition and pecking. It is important to note that ground feeding can increase the risk of contracting diseases or parasites through contact with droppings from contaminated birds, so if that is your practice, always try to feed in the cleanest area available. If you ground-feed, you can scatter feed once early in the day and again later in the day before everyone goes up to roost. Try not to put out more feed than your turkeys will eat by the end of the day. Excess feed left on the ground at night will attract rodents.

Although the better feed conversion ratio of broad-breasted turkeys is a benefit for those raising Thanksgiving birds, the flip side is that to keep them beyond Thanksgiving, you may need to restrict their diet to reduce the risk of obesity and its associated health challenges. Adult broad-breasted turkeys will eat a lot of feed if it is freely available, and they typically eat quite a bit more than heritage varieties. The longer you keep a broad-breasted turkey alive past maturity, the more the cost savings of their greater feed efficiency diminishes.

Water

A CLEAN SOURCE OF DRINKING WATER is an important element of overall nutrition and should be provided year-round. Although free-ranging turkeys can find ponds or streams to drink from, allowing them to rely on natural water sources is not the best practice because of the risk of turkeys picking up diseases from wild birds, in particular water-birds. Provide waterers close to home, and change the water as needed to keep it fresh. I prefer to use the 3- to 5-gallon plastic waterers that have a deeper reservoir at the base than the galvanized waterers.

An A-frame structure keeps water clean by preventing birds from roosting on the waterer.

Trying New Treats

Turkeys tend to be very suspicious of new things, and this includes treats. Unlike chickens, who are often more than happy to help their humans take care of any surplus garden harvest, my turkeys don't have particularly adventurous palates. They prefer to stick with black oil sunflower seeds, scratch grains, and dried black soldier fly larvae (grubs). My turkeys have learned from the chickens that scrambled or hard-boiled eggs are also a delicious treat, but anytime I try to give them a pumpkin or something new from the garden, they are more likely to ignore it than appreciate it.

The turkeys know that the big red cup means it's time for black oil sunflower seeds, a favorite treat.

Supplements

GRIT. Many birds, including turkeys, need grit to properly digest their food. Grit is nondigestible material, usually small pebbles, that the birds swallow; it sits in the gizzard to grind food. Whether you are feeding your turkeys a whole-grain feed, giving them scratch grains, or allowing them to forage on pasture, offer them free-choice grit so they can pick up as much as they need. Free-ranging turkeys can find some of the grit they need but may not find enough or the right size. Grit is available for each stage of development: chick, grower, adult.

CALCIUM. Eggshells are mostly calcium, which laying hens need a lot of to produce their big, beautiful eggs. Even if a layer feed includes calcium, offer a free-choice source—either crushed oyster shells or crushed limestone. The need for calcium varies among hens, and some hens may need more than what they can get from their feed. In addition, older hens need increasingly more calcium, not only to ensure that they continue to lay hard-shelled eggs but also for other bodily functions, proper bone strength, and overall health.

OTHER SUPPLEMENTS. Turkeys are generally very hardy and healthy birds, but I still like to provide them with powdered vitamins and electrolytes in their water several times a year. I try to do this a few days in advance of stressful weather conditions. If the forecast is calling for especially hot temperatures or especially cold winter weather, these are both good times to provide them with vitamins in their water to give their immune systems a little boost.

Different sizes of grit, from chick to adult

Pasture Rotation

THIS MAY BE STATING THE OBVIOUS, but turkeys are big, and they poop a lot! Manure management is important because diseases and parasites are often transmitted through manure. Manure management is less of an issue for turkeys that free-range, but since turkeys will wander a very long distance if left to their own devices, you may consider portable range fencing to keep them closer to home, even if you do have a sizable ranging area. For turkeys kept in fenced areas, occasional rotation is a good idea.

One of the nice things about turkeys is that they tend to do a lot less damage to grass and vegetation than chickens do. Turkeys like to eat the tips off growing vegetation, and they will peck around for goodies on the ground. Although they do scratch the ground as chickens do, turkeys tend not to dig as many holes nor kill vegetation quite as readily. It is fairly easy to keep a turkey pasture vegetated and looking green with some occasional attention to pasture rotation. Feeders and waterers should also be occasionally moved to a new location if manure accumulates around them. Or you can put down coarse wood shavings over soiled areas to keep the ground cleaner.

For turkeys, the grass is usually just as green inside the fence as outside.

The frequency with which you rotate your fences depends on the size of your pasture, the number of birds, and how well the grasses and forbs regenerate in different climates and during different seasons. Our portable poultry netting with step-in posts works well for most of the year, except during the driest months of the summer when our clay soils become rock hard, making it nearly impossible to move the step-in posts. For this reason, we always rotate the fence before the rain stops in the summer, while the ground is still soft enough to move the posts.

We enlarge the size of the fenced pasture in the summer to give them plenty of fresh areas to graze, and the fence stays in place throughout the summer. After the fall rains arrive, we move the fence again and reduce the size of the pasture to keep them a bit closer to home during the rainy season, when they tend to forage less. In the spring when the weather improves, we enlarge the size of the pasture again.

Dust Bathing

DUST BATHING IS AN IMPORTANT PART of a turkey's grooming routine. Turkeys rely on dust bathing to discourage external parasites, such as mites and lice, and to keep their skin and feathers in good condition. They seek out patches of bare ground and scratch and dig to loosen up the dirt and form a bit of a wallow, after which they lie down in the fresh dirt, roll around, and use their wings to throw dust up onto their body. This activity is often accompanied by happy purrs of contentment from the turkey hens and is always fun to observe.

Dust bathing is a year-round activity, so make sure your turkeys have access to a suitable dry area. Both hens and toms engage in dust bathing, although toms tend to be more preoccupied with showing off than dust bathing, and I don't see my toms dust bathing very often. For this reason, it's a good idea to occasionally examine your turkeys for external parasites. When your tom is busily strutting his stuff, take advantage of the opportunity to give his backside a quick look to make sure he is parasite-free. (See Chapter 8 for more about parasites.)

Turkeys need access to a suitable dust-bathing spot all year long.

Seasonal Care Considerations

EXTREMES OF HEAT AND COLD present challenges for turkeys that live outdoors, even if they have adequate shelter. Here are a few issues to be aware of to keep everyone healthy during difficult weather conditions.

Summer

Turkeys can easily become overheated in the summer. When the temperature is above 80°F (27°C), you may notice signs of heat stress such as wings being held away from bodies, open beaks, and panting. Providing fresh, cool water is especially important during hot weather. Place a source of clean water in the shady areas where your turkeys take respite during the hottest part of the day. Multiple sources of water are better. Even though I always have two large stationary waterers in the turkey yard, I find that when the birds are hunkered down, they aren't willing to walk very far to get a drink. They also may not drink water that has been warmed by the sun. Adding frozen berries to a shallow pan of water or offering them slices of chilled watermelon may encourage your turkeys to stay hydrated on especially hot days.

On days when overheating is of extra concern, I bring out several of the heavy-duty shallow feed pans filled with fresh water and sit them right next to where the turkeys are resting in the shade. In addition to

Whenever my broody turkeys take a break from their nest, they go straight for a cooling footbath.

drinking more readily from fresh, cool water that is nearby, they also like to stand in the pans of water, which helps them cool down a bit. It's a good idea to check the waterers a few times a day, to make sure that the water is still clean and hasn't been fouled or tipped over and to provide fresh water to encourage everyone to stay hydrated.

Another good measure you can take to help your turkeys stay cooler on hot days is to wet down the ground in the shady areas where they like to hang out. Do this first thing in the morning before the heat of the day. This will keep the ground a few degrees cooler and in turn will help the turkeys stay a little cooler when they lie on the dampened earth. You can also water down their dust-bathing areas—bathing in dampened dirt will help them regulate their body temperature.

Winter

Winter can be one of the most frustrating times for turkey owners, because no matter how hard we try to keep our turkeys sheltered and dry, there are often a few independent spirits that will shun our best attempts and insist on sleeping outside in the rain and snow.

Winter Hardiness

Turkeys are extremely hardy birds, and they are perfectly happy spending time out in the open during what we would consider cold and unpleasant conditions. Turkeys will often continue to forage during light rain or snow, and they prefer to stay active as long as conditions aren't too miserable. If conditions get extremely rainy or snowy, turkeys will eventually exhibit some common sense and seek cover under a roof or on their favorite roost.

Not only is it common to see turkeys outside going about their usual business on a cold winter day, but it's common for turkeys to want to sleep outside even on stormy nights. Their body mass, large wings, water-resistant outer feathers, and many layers of downy under feathers keep them warm through the night as well as on the coldest of days. Even if their outer feathers become wet from the weather, if you put your hands beneath their wings into their under feathers, you can reassure yourself that they are warm and dry despite having a soggy outer appearance.

Even heavy snow won't deter turkeys from enjoying the open air.

Turkeys can handle sleeping outdoors in cold and wet weather if their feathers are allowed to dry out occasionally. Being cold and wet for a prolonged period puts them at risk for developing pneumonia or other respiratory ailments. During breaks in bad weather, turkeys will be busily preening and spreading their wing and tail feathers to dry.

Winter weather is not the only potential danger for turkeys that roost outdoors—predators are extra hungry in the winter and more likely to come around looking for easy pickings. If you can't convince your turkeys to sleep indoors, consider taking extra steps to keep them safe such as putting an electric fence around their roosting area or getting a livestock guardian dog.

If you have misbehaving turkeys, you might want to consider adding one or more of the shelters discussed in Chapter 4, especially if there is no roof over their favorite nighttime roosting spot. It's best to plan ahead if you want to make changes to their sleeping accommodations for the winter, as any changes will likely be viewed with suspicion. Give your flock at least a few weeks to get used to any changes before the worst of the winter weather arrives.

For the more cooperative turkeys that do sleep in an enclosure, having good airflow and ventilation is very important during the winter. Poultry that spend more time inside during the winter need their bedding replaced more often to keep it dry and clean. Being indoors means they also kick up a lot of dust, which can cause issues for their sensitive respiratory systems. The cleaner and dryer you can keep the shelter, the better.

Bedding tends to break down into small particles over the course of several months. Depending on your setup and the number of turkeys sleeping in or under a nighttime shelter, I recommend scooping or raking up droppings every week or so, and adding more bedding as needed. (I use coarse wood shavings or chips.) Regardless of your cleaning schedule, it's a good idea to do an extra clean-out before winter and replace any damp, moldy, or potentially dusty bedding with a thick layer of fresh bedding.

Providing extra calories helps your turkeys stay healthy in cold weather. I throw out extra scratch at treat time, and a couple of hours before roosting time, I bring them a generous portion of scratch mixed with black oil sunflower seeds. Burning the extra calories helps them keep a bit warmer at night.

SNOW TURKEYS

We woke up on Christmas Eve to fresh snow. I wondered if the turkeys would not be quite as excited to see the snow as I was. Our chickens can choose to stay inside and have their morning food and water in the comfort of the coop without venturing into the snow. The turkeys, however, do not have the same luxury because they prefer to sleep outside on their high roost no matter what the weather. On this morning, they woke up to snow not only covering the ground but also covering them!

In the morning when I go out to feed, the turkeys are usually roaming the pasture. On their first snow day, they were still on their roost, hesitant to fly down. I could see that Ringo's breast feathers were soaking wet and he was shivering. Prudence and Eleanor looked like they were ready to get out of the snow, too. I felt sorry for them and thought they would be happier if they were down off their roost and on the covered chicken coop porch, where they could dry off.

The turkey roost is 6 feet off the ground, so I got a step stool to try to get them down. It was pretty awkward trying to wrangle the turkeys without getting smacked in the face! I managed to coax Prudence and Ringo to step onto my arm, but I had to grab Eleanor rather unceremoniously since she is less tame. I finally got them all down and onto the coop porch for breakfast. Prudence and Eleanor spent much of the day preening and drying out on the porch while Ringo explored the snowy pasture.

I hoped that the snow would encourage the turkeys to spend the night in the coop, but by the afternoon Prudence and Eleanor had begun exploring the snow, and the turkeys spent the night on their outdoor roost as usual. The next morning there was still snow on the ground, but this time the turkeys flew down off their roost and ran to greet me for breakfast as if the snow was no big deal, although the chickens had not come to the same conclusion and were still waiting for me in the coop!

CHAPTER 6

Raising Poults from Day 1

It may surprise you to learn that poults are not as robust as chicks and need plenty of attention in the beginning to get off to a healthy start. If you simply put the young birds in the brooder and let them fend for themselves the first few days with little supervision, you may lose a few poults. But with some extra TLC in the critical first week, many people experience no losses.

Poults tend to be more easily chilled than chicks, so pay special attention to keeping them warm while they are being transported and get them into their brooder as soon as possible. Before bringing them home, set up the brooder with bedding, food, water, and a heat source so the poults can immediately start settling into their new home.

These little cuties are hard to resist, but give them a couple of days to warm up, rest, and eat and drink before you start handling and playing with them. When you do start to handle the poults, keep it to a minimum—a few minutes at a time—to avoid stressing or chilling them. Always supervise young children when they are handling the poults.

Differences in Brooding Poults versus Chicks

POULTS	CHICKS
More sensitive to chilling and stress	Hardier when newly hatched
Need brooder temperature started at 100°F (38°C)	Need brooder temperature started at 95°F (35°C)
Need help learning to eat and drink	Quicker to find food and water
Need 30 percent protein starter feed	Need 18–20 percent protein starter feed

poult

chick

Where to Purchase Poults

THERE WAS A TIME WHEN BROAD-BREASTED TURKEYS were the most common—or only—type of poults sold by farm and feed stores. Nowadays it is easier to find at least a few varieties of heritage poults for sale as well. If you can't find the variety of heritage turkey you are looking for at your local feed store, try a hatchery that offers a wide variety of poults available for shipping.

Feed Store

A feed store, if there is one near you, may be the most convenient location to buy your poults. Poults are sold as straight-run or unsexed, meaning they have not been sorted by sex, and you are just as likely to get a male as a female. Poults are typically available a little later in the spring than chicks. It's a good idea to wait to buy poults until a couple of days after they have arrived at the store. This gives them a little time to recover from their stressful journey and to get some nourishment.

As with any in-person poultry purchase, you should observe the poults in their brooder and select ones that appear healthy—they should be active, eating, and drinking. In any shipment of young poultry, there are usually a few off in the corner looking hunched up and generally unwell. As tempting as it may be to try to help the underdog, there is often nothing you can do to help a little one that may have suffered too much stress during shipping or may have an internal defect. Young poults are fragile, and losses are not uncommon in the first few days of life. My recommendation is to save yourself the heartache of a lost cause and pick from the healthiest-looking poults to give your flock the best chance of getting off to a good start.

Hatchery

Hatcheries are a good option if you have a specific heritage turkey variety in mind and it is not available at your local feed store. Hatcheries often set a minimum purchase number to ensure that the poults stay as warm as possible during shipping. All poults are shipped as straight-run, and it is not possible to select males or females. Buying mail-order poultry can be a stressful experience, as you watch the online status of your shipment nervously and keep your fingers crossed that there are no delays, which are usually fatal for the young birds.

Of course, plenty of people order by mail with good results. I have purchased mail-order poultry twice and although both times the shipments arrived quickly, there were dead birds in each shipment. If you don't do well with these types of anxiety-inducing situations, buying poults from your local feed store may be a better option.

Private Seller

When buying poultry from an individual, a farm, or an auction, it's important to understand the potential risks involved with bringing birds in to your flock from an outside source. Buying poultry from a farm that participates in the National Poultry Improvement Plan (NPIP), which is designed to improve flock health and to ensure the sale of disease-free poultry nationwide, is one way to minimize the risk. Not all farms have NPIP certification or have the strict biosecurity measures in place that hatcheries do.

The downside of auctions is that you don't have the opportunity to see the conditions in which the poultry were raised. If you are buying birds from a private seller or a farm, ask to see where the birds have been raised so you can see for yourself if their environment is clean and if their flock mates appear healthy. I once visited a local farm that I found online when I was shopping for adult turkeys. When I saw the unsanitary conditions, I knew there was no way I was taking a chance on bringing home potentially unhealthy birds. I quickly got back in my car, and I sterilized my shoes and washed my clothes as soon as I got home. You really can't be too careful when it comes to avoiding the risk of bringing home a contagious disease to your flock.

Two-day-old mail-order poults

If you do decide to buy poults or mature turkeys from a private seller, farm, or auction, quarantine the new birds from any other poultry on your farm for a few weeks to ensure that the new additions are healthy before introducing them to your flock.

Setting Up the Brooder

A BROODER IS A SAFE, enclosed, heated space for raising young poultry when they aren't raised by a mother hen. This is where your poults will spend the first few weeks of their lives. There are a variety of options for a brooder. Poults grow quickly, and they also start to fly early on, so a setup for chicks may not be suitable for poults. Smaller brooders made from plastic storage bins or cardboard boxes or low-sided kiddie pools are not ideal. I use a wooden brooder box built with untreated plywood. A 2 × 4-foot box whose sides are at least 2 feet high is roomy enough for up to four poults. A stock tank is another good tall-sided option. The 2-foot-high sides will keep the poults from jumping out for the first couple of weeks, but after that they require larger accommodations. Disinfect any container that has been used for raising poultry before using it for poults.

Depending on the number of poults you plan to raise, you may opt to start with a larger brooder in a secure, sheltered area such as a garage or barn. Take special care to protect the brooder from drafts and to ensure that it will stay warm, considering the greater variation in temperature compared to a well-insulated location. You will also need to take extra precautions to keep the poults safe from predators, including barn cats and farm dogs, until you can properly introduce those animals to the newest members of the farm family.

Set up the brooder with feed and water and turn on the heat lamp before bringing home your poults.

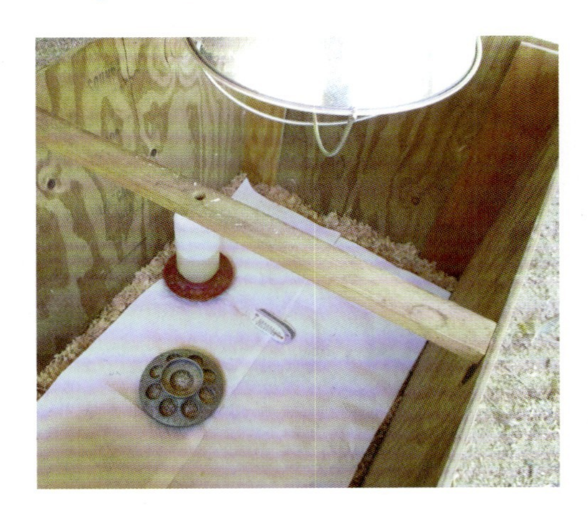

Indoor Brooders and Toxic Fumes

If your brooder is near the kitchen, you must provide excellent ventilation. Letting a Teflon-coated pan overheat on a burner or using chemical oven cleaners can generate toxic fumes that can be lethal to young poultry. Better yet, don't cook with Teflon or clean your oven while you have a brooder in the house.

Whichever option you choose, cover the top of the brooder to keep poults from flying out. An inquisitive poult that flies out of the brooder and onto the floor won't be able to find its way back into the brooder and will soon become chilled, to its detriment. Be sure that the brooder cover allows for air circulation. Options include ½-inch hardware cloth, window screening, or plastic garden fence with very small openings.

The cover must be weighed down or affixed to the sides of the brooder so that it can't be easily dislodged. Make sure the material used for the brooder cover is not something that curious poults can fly up into and get caught. I would not recommend using chicken wire or a woven netting for this reason. Similarly, don't ever use chicken wire for the walls of a brooder enclosure; it's far too easy for curious poults to stick their heads through and get stuck, potentially injuring themselves.

BROODER CHECKLIST

- Brooder with cover
- Bedding material
- Heat source
- Thermometer
- Feeder and waterer
- 30 percent protein starter feed
- Vitamins and electrolytes

Brooder Size and Space per Poult

For the first couple of weeks, give your poults a minimum of ½ square foot of floor space each—1 square foot is even better. Poults grow rapidly, so it makes sense to build a brooder with plenty of room. If you are starting with just a few poults, you can block off part of the brooder for the first week to make sure everyone stays in the warm side of the brooder. After the first week or so, the poults will be quite active, so the bigger the space you can give them, the better. More space will keep them from becoming bored, which can lead to feather pulling and toe pecking.

Plan to provide poults with at least 2 square feet of floor space from the time they are a couple of weeks old until they are eight weeks of age. I know some sources provide space recommendations of 1 square foot per poult up to six or even eight weeks of age. Based on my experience, this is not sufficient to prevent boredom and undesirable pecking behaviors. The more space they have, the less likely they are to start displaying cannibalistic behaviors. If you see this type of behavior, move them into a larger brooder space as soon as possible.

Poults can fly up to high roosts at a young age.

Within a couple of weeks, the poults will surprise you with their flying ability and propensity for exploration. This is a good time to add a low roost. It should be 1 to 2 inches wide and about 6 inches tall. When the brooder is no longer big enough to contain your rapidly growing poults, or you get tired of the dust that goes with having an indoor brooder, you can transition your poults outside to a larger, but still sheltered, outdoor brooder. If your poults will have their own coop (one not currently occupied by other birds), you can move them into their coop—see Poult Housing Needs, page 110. Remember that poults need a heat lamp until they are fully feathered and outdoor temperatures are no cooler than 60° to 65°F (16° to 18°C).

Bedding Material

THERE ARE A VARIETY OF OPTIONS for bedding material in the brooder. Some have been in use for generations and others are relatively new. I will cover the basics here, and as with so many decisions related to raising animals, you should experiment to determine the method that works best for your setup and budget. Some of the most common bedding options include pine shavings, straw, wood pellets, and hemp. All these materials are compostable.

Never use newspaper because it is slippery for young poults. They can easily dislocate a leg on the slick surface, leading to spraddle leg. Sawdust is not a good choice either because the particles are too fine and can injure poults' sensitive respiratory systems.

PINE SHAVINGS are a popular choice—they are relatively cheap, they do a good job absorbing moisture from droppings and keeping the brooder dry, and they are readily available. They are also lightweight and easy to work with. Coarse shavings are better than fine shavings, as fine shavings tend to be dustier. Do not use cedar shavings. Cedar contains highly aromatic essential oils that can damage respiratory systems in enclosed spaces.

STRAW is an affordable and readily available option that has been used by generations of poultry keepers. Some drawbacks are that it is not very absorbent, it tends to mat down, and it isn't as easy to keep clean as other options. Some people have concerns about poults eating straw, which may result in crop impaction. If this worries you, there are plenty of potentially safer options.

WOOD PELLETS, used as bedding in livestock stalls, are inexpensive and readily available. The compressed pellets are also very absorbent and break down into sawdust-like matter when wet.

HEMP is a more recent bedding option. Once hard to find, it is increasingly available, although the price remains high compared to other materials. It is made from the stalks of industrial hemp plants and is reportedly less dusty and more absorbent than some other bedding materials.

I prefer to use pine shavings in the brooder. They check a lot of boxes, including low cost, availability, and ease of cleaning. One of the downsides to pine shavings, especially smaller shavings, is that they generate dust when all those little feet get to scratching around in the brooder. It doesn't take long before you will notice a fine coating of dust has wafted through your house and settled on all the furniture. In my opinion, the enjoyment to be had from being able to easily check on the poults by keeping them inside the house for the first few weeks is well worth a little extra housecleaning.

If you are using pine shavings, straw, or hemp, put down a 3- to 4-inch layer in the brooder. For the first several days, put paper towels over the bedding material where you've set up the heat lamp and food and water, which is where the poults will spend most of their time. The paper towels allow the poults to better see and identify their food. You want those first few meals to be starter feed, not bedding! Paper towels also allow you to keep the brooder drier and more sanitary in the first few days, as you can easily swap out soiled ones. After the poults have figured out how to eat and start scratching around in the bedding and running around the brooder, you can remove the paper towels.

A layer of paper towels over the shavings helps the poults see their food better and makes it easier to clean the brooder.

Heating Options

THERE ARE TWO MAIN TYPES OF HEAT SOURCES for use in the brooder: heat lamps and brooder plates. It is important to have the brooder warm before you bring your poults home. Turn on the heat source the day before your poults arrive to get the brooder up to temperature, and make any necessary adjustments. It's helpful to place a thermometer a few inches above the bedding so that you can see what the temperature is at poult height. If you are using a heat lamp, hang it over one side of the brooder to create a warm side and a cooler side in the brooder. This will allow poults to self-regulate their body temperature.

The temperature of the brooder should be about 5°F (2.7°C) warmer than when brooding chicks. For poults up to a week old, the temperature should be 100°F (38°C) in the bottom of the brooder. Reduce the temperature 5°F (2.7°C) per week and continue to provide supplemental heat until the poults are fully feathered, which is typically by seven or eight weeks of age.

To prevent poults from flying onto the cord of the heat lamp and potentially knocking it down, be sure that the cord does not dangle over the brooder.

Heat Lamp

A heat lamp comprising a red 250-watt bulb, a metal reflector, and a protective guard/cover is the least expensive option. The heat lamp can be hung or clamped above the brooder. Start by positioning the heat lamp approximately 18 inches above the floor of the brooder, and place a thermometer at poult height to monitor the temperature. You'll adjust the height of the heat lamp regularly to control the temperature in the brooder; keep this in mind when setting up the heat lamp. One way to do this is to securely attach a 5-foot length of 1×2 board to the side of the brooder. You can then securely clamp the heat lamp to this board and easily adjust its height. Another option is to hang the heat lamp on a chain from a hook in the ceiling.

Due to its high temperature, a heat lamp poses some safety risks. The first is the potential for burns if you, your children, or pets brush against it. The second is the potential for fire if the heat lamp should fall into flammable bedding material. Secure the heat lamp tightly to something sturdy and make sure the cord does not dangle over the brooder. You don't want the poults to fly up onto the cord and dislodge the lamp.

To determine if the temperature is comfortable for the poults, watch how they situate themselves in the brooder with respect to the heat lamp.

When poults are cold, they will huddle together under the heat lamp. Poults that are too warm will spread out and move away from the heat lamp. When they are evenly spaced throughout the brooder, you can be assured that the temperature is just right.

To reduce the temperature in the brooder, you may need to raise the heat lamp quite a bit. I've raised mine by as much as a foot at a time when trying to lower the temperature by 5°F (2.7°C). Again, keep this in mind when arranging your setup and planning for how you will adjust the heat lamp because you may need to adjust it more often than you think. On days when the weather is especially warm, you may need to turn off the heat lamp for a few hours if the poults seem uncomfortable, but set an alarm to remind yourself to turn the heat lamp back on after the heat of the day subsides. On particularly cold days, the poults may appreciate the extra warmth if you lower the heat lamp by several inches.

Using the Right Type of Bulb

It is very important to use the right type of bulb for the heat lamp. Do NOT use shatter-resistant or safety light bulbs. These bulbs often have a coating of Teflon (polytetrafluoroethylene or PTFE), and when the bulb heats up, it can release fumes that are toxic to poultry and can quickly cause death. It's also important to use a red bulb and not a white one. The brighter light cast by a white bulb can disrupt the poults' sleep cycle and can lead to aggression and feather picking. The light cast by a red bulb makes it a little harder for the poults to see blood in the event there are

Monitoring Brooder Temperature

How the poults are behaving is the best indicator of whether the brooder temperature is too cool, too warm, or just right. If the poults are moving around, peeping, eating, and sleeping, then the temperature is fine. The behaviors listed below indicate that the temperature needs adjusting.

- Huddling together under heat lamp—indicates chilling
- Cold feet—indicates chilling
- Huddled at side/in corners of brooder—indicates overheating
- Panting—indicates overheating
- Loud, distressed peeping—indicates chilling or overheating (also possibly hunger/thirst or sickness)

any pecking injuries, and so it is helpful in reducing cannibalistic behavior. It's a good idea to have a spare heat lamp bulb on hand because when one burns out, it will inevitably be at the most inopportune moment or when the feed store is closed.

If you find that your poults are getting aggressive and pecking at each other even with a red bulb, it is a sign that they are bored and they need more space in the brooder. Consider moving them to a bigger space to help reduce these undesirable pecking behaviors.

Brooder Plate

A brooder plate is designed to mimic the natural environment of a momma hen brooding her poults underneath her feathers. It is a flat, rectangular plate that the poults stand underneath to stay warm. These plate-style brooders commonly include adjustable legs that can be raised as the poults grow larger. The height of the plate should be set so that poults can stand comfortably underneath it without their heads touching the plate. There are also brooder plates designed to hang.

Brooder plates are safer in terms of being cooler to the touch and avoiding the risk of burns, and they also pose much less risk of fire. The downsides are that they are more expensive to purchase than heat lamps, and depending on the number of poults you are raising, you may need more than one brooder plate to keep everyone warm, especially as the poults grow larger or if you live in a colder climate.

Cleaning the Brooder

IT IS VERY IMPORTANT TO KEEP THE BROODER CLEAN AND DRY. Wet bedding should be removed immediately, but how often you need to clean up droppings and change all the bedding will depend on the size of your brooder and the number of poults. If you have a lot of poults in a small space, you may need to clean up a couple of times a day. With fewer poults or a larger space, you may be able to get by with a light daily cleaning and a weekly full bedding change.

The risk of leaving moist bedding and droppings for too long is that they provide the perfect environment for coccidiosis to develop and spread. This intestinal infection can be deadly in a very short time if it is not caught and treated quickly (see Chapter 8). It's best to practice regular cleaning of the brooder and not endanger the health of your poults by inviting in a disease that is entirely preventable with good hygiene.

Hanging feeders and fount-style waterers

Feeders and Waterers

POULTS QUICKLY OUTGROW THE FEEDERS AND WATERERS sold for chicks, so you may want to skip the smallest sizes of feeders and waterers and start with something a little bigger that still fits comfortably in the brooder. There are metal and plastic options for feeders and waterers, although the traditional metal versions are rapidly being replaced by plastic versions.

The advantage of metal feeders and waterers is that they last longer, although they do eventually rust. The plastic versions are lightweight and easy to clean, but they tend to be less durable over the long term. There are several designs of feeders and waterers, each with pros and cons as discussed below.

Avoid putting a feeder or waterer directly under a heat lamp; it may discourage poults from eating and drinking if the brooder temperature is too warm in that location, and warm water is not as appealing as room-temperature water.

Feeders

It seems that every few years manufacturers introduce a new style of feeder that attempts to minimize feed waste and avoid feed contamination. After all, it's incredibly annoying to see the feeder you just cleaned already filled with bedding or, worse, poop! Some of the newer styles of feeders may not be as easy for young poultry to use or may not allow as many poults to feed at once, and therefore they may not be as suitable for young poults as they are for older birds. The main styles of feeders traditionally used for very young poults are:

- Round base and mason jar
- Rectangular trough
- Hanging feeder

A seven-week-old poult at trough feeder

When I first started keeping poultry, I used the round mason jar style of feeder in the brooder and found refilling it to be messy and tedious. I also tried a hanging feeder suspended from a bar across the top of the brooder. This made it simple to raise the feeder as the poults grew larger, but I couldn't easily relocate the feeder to a different area of the brooder like I could with one that sat on the ground. Now I use the rectangular trough-style feeder, which is convenient to refill and easy to move around the brooder if I decide to rearrange the setup. Because it's long, it allows plenty of room for everyone to access the feed, and it also works well as the poults grow larger.

Teach poults to find their food by tapping your finger on the feeder.

Poults tend to kick bedding material into the feeder if it is placed directly on the floor, so it's a good idea to raise the feeder an inch or two above the bedding. If you are handy, you can build a wooden stand or platform for holding your feeder. It's also perfectly fine to simply use a piece of scrap lumber or a brick; just be sure whatever you use will not tip over easily when the poults inevitably jump up onto it or onto the feeder.

Set the feeder a bit higher than the bedding to keep the feed clean but not so high that the poults can't access it easily. You will likely need to adjust the height and location of the feeder as the growing poults become more active.

Poults can take a little longer than chicks to learn where their food and water are. It's a good idea to show them where the food is as soon as they are placed into the brooder. You can do this by tapping your finger repeatedly in the feeder and making a soft encouraging sound. (I like to say, "Here, turk, turk.") In this way you will be simulating the behavior of a momma hen as she calls her babies over to show them what to eat.

Chicks as Teachers

Some people are big proponents of adding a couple of chicks to the brooder right alongside the poults. Chicks tend to be a bit quicker at learning where the food and water are, and the poults will learn how to eat and drink by following their example.

To help teach them to eat, for the first few days you can also place one or two small dishes of feed (I like to use plastic lids) in the brooder or sprinkle a few pieces of feed onto the paper towels that are covering the bottom of the brooder. If you use white paper towels, the color of the feed will contrast nicely with the white, making it easy to see. Tap at the feed multiple times a day to encourage the poults to eat. Be sure to watch the poults regularly to make certain they have all figured out the feeder and waterer and your setup offers enough room for everyone to eat and drink.

Waterers

Instead of using the smallest 1-quart waterers typically sold for baby poultry, I like to start off with a 1-gallon waterer. I prefer the style of waterer with a reservoir around the base. This gives everyone easy access to the water, and I find that the greater weight of the 1-gallon waterer prevents it from tipping over. Poults can easily kick bedding material and poop into the waterer if it is placed on the floor of the brooder, so it should be raised an inch or two above the height of the bedding in the same way as the feeder.

When using a fount-style waterer, it's a good idea to place marbles into the base of the waterer to prevent a poult from falling in and drowning. Keep the marbles in the waterer for the first week or two, until the poults get the hang of things and are a bit stronger and steadier on their feet. You will find that no matter how you try to keep the waterer from getting contaminated, the active poults will make a mess of it several times a day. Be sure to change out the water whenever it's dirty.

Another option is a nipple-style waterer. There are different designs of nipple waterers that you can buy or make yourself by installing nipple attachments on either the side or the bottom of a plastic container, bucket, or PVC pipe. A nipple waterer works by means of a metal pin at the end, which releases a few drops of water when it is pecked. This is a closed system, and the water stays cleaner because shavings and poop can't be kicked into the container. It reduces the time spent cleaning the waterer and the risk of disease caused by drinking dirty water. If the system is functioning properly without leaking, water won't spill onto the bedding, which is another advantage.

Potential downsides of this style of waterer are that it takes a little more learning to figure out how to use it, and it doesn't allow as many poults to drink at one time. If you decide to use a nipple waterer system, watch your poults closely to make sure that everyone has figured out how to get the water they need.

I have used nipple waterers for adult chickens, but I don't have any experience using them for poults or chicks. I want to make it as simple as possible for my young birds to get enough to eat and drink, so even if it makes a little more work for me keeping feeders and waterers clean, I prefer the rectangular trough-style feeder and fount-style waterer. These designs are easy for poults to use and have been around for generations.

Show the poults where the water is and encourage them to drink multiple times a day for the first couple of days until you are certain that everyone has figured it out. When placing poults into the brooder, it is a good idea to gently dip their beak into the water to show them where it is. Putting a few glass beads or marbles (large enough that they cannot be eaten) into the base of the waterer encourages the curious poults to explore the waterer. They will be attracted to the shiny objects, and in pecking at the beads or marbles, they will inevitably get a little taste of water, which reinforces that it's good to drink.

A layer of shiny marbles in the waterer encourages curious poults to drink and prevents very young ones from submerging their faces and possibly drowning.

Feeding and Nutrition

THE NUTRITIONAL REQUIREMENTS FOR POULTS differ from those of chicks in one key way: protein. Poults grow more quickly than chicks do and therefore need a higher-protein feed in their first four months of development. This is true for both heritage and broad-breasted varieties. Feeding a starter feed with insufficient protein content can lead to developmental problems and even the death of young poults.

Starter and Grower Feed

For the first eight weeks or so, poults should eat a high-protein turkey or game-bird starter. Give them a 30 percent protein starter, or as close to 30 percent as you can find. Check the brand of feed you are using to determine when the manufacturer recommends transitioning from starter feed to developer feed. Different brands vary slightly in their formulations and feeding recommendations, but typically poults are switched to a lower-protein (approximately 20 percent) developer/grower feed between six and eight weeks of age. They can be moved to a layer or maintenance feed at the age of four months.

If you are feeding a crumble-type or pelleted starter feed that does not have any whole grains in it, there is no need to offer grit in the first few days. However, as soon as you introduce solid food to the poults, such as dried grubs or mealworms, grains, or greens as treats, be sure to also provide grit (tiny rocks or pebbles). Since poultry do not have teeth, they use grit in their gizzard to grind up their food. Sprinkle a small amount of grit on top of their feed when

Starter feed sprinkled with grit

Feeding Chart

AGE	TYPE OF FEED	PERCENT PROTEIN
Up to 8 weeks	Starter	30
8–16 weeks	Developer/grower	20
16 weeks on	Grower, maintenance, or layer	16–20

you top up the feeder every day so they can build up the grit they need in their gizzard.

Nutritional Supplements

To get young poults off to a healthy start, it's a good idea to put powdered vitamins and electrolytes in their water for the first week. They may be dehydrated after being shipped or stressed by settling into a new home, and vitamins and electrolytes can give them a helpful extra boost. Be sure to follow the package instructions and don't over-administer vitamins and electrolytes. Excessive amounts of vitamins can potentially be harmful and cause nutritional imbalance, so it's not a good idea to administer vitamins and electrolytes constantly.

PROBIOTIC SUPPLEMENTS AND HERBS. Probiotics are beneficial microorganisms that contribute to a healthy gastrointestinal system. They may be helpful when administered in times of stress. Herbal supplements are another trend that has gained in popularity in recent years. Various herbs are reported to be beneficial for boosting the immune system or contributing to respiratory health. Opinions vary on whether it is helpful to give probiotics or herbs to poults. I have never used these types of supplements, although farms seeking organic certification for their poultry rely on them. As with so many things in raising poultry, do your own research and do what feels right for you and your flock.

APPLE CIDER VINEGAR. In my early years of keeping chickens, I routinely added apple cider vinegar (ACV) to the chickens' water because I had read on several chicken blogs that it was good for them, and I wanted to do everything I could to keep my ladies healthy. After a few years, as my flock grew along with my to-do list, I began to question the value of this practice.

I tried to research the pros and cons of using ACV online and found all kinds of opinions either for or against it. However, most of these opinions appeared to be based on anecdotal experience, and I couldn't find any concrete evidence based on scientific studies of backyard flocks that adding ACV to their water had a tangible benefit. While there may be some consensus that it can contribute to gut health, there are also many false or unsubstantiated claims that ACV is a miracle cure for all manner of parasites and ailments. I stopped using ACV several years ago and haven't noticed a difference in the overall health of my birds.

Treats

One of the great joys of raising poultry is seeing their excited little faces at treat time and having them rush toward you as if you are their favorite person in all the world! It's okay if you want to keep on thinking that (I do), but everyone knows that it's all about the treats.

For the first couple of weeks, it's best to keep poults focused on getting maximum nutrition from their well-formulated starter feed. I start giving treats sparingly once they are a couple of weeks old. Treats are a great way to start training the poults to come when called, but treats should make up no more than 10 percent of their daily diet.

Poult Housing Needs

THE AGE AT WHICH POULTS SHOULD BE TRANSITIONED from the brooder to their permanent housing will be determined by the size of your brooder and the number of poults you have. In general, if you are using some sort of container as a brooder (as compared to a large brooder room), poults will outgrow their brooder rapidly and will be ready for more space by the time they are three weeks old. At this time they can be moved into new purpose-built turkey housing or a partitioned-off section of barn until they can join the rest of the flock at about eight weeks of age. Poults should be confined inside their new housing day and night for several weeks to reinforce the habit of sleeping indoors.

Poults will outgrow a brooder at about three weeks of age.

The natural curiosity of poults can get them into trouble sometimes, so make sure that there are no gaps in their housing that they might squeeze into and get trapped, or push through entirely and be unable to get back in. When moving poults to outdoor housing, remember that a heat source must be available until the poults are fully feathered at seven to eight weeks of age. In addition, young poults have very delicate immune systems, so they should not be released into areas that adult birds are using until they are eight weeks old.

Transitioning Poults Outside

Mortality of poults can occur during the period when they are first transitioned outdoors. This can be due to disease from soilborne pathogens, exposure to weather, or predators. It's a good idea to keep poults off the ground, especially if adult birds have been using the area, until the poults are eight weeks old. Poults that are hand-raised by humans (as compared to poults raised by a momma turkey) seem to be especially vulnerable to contracting diseases from the ground. On the other hand, poults that are raised by their mother are more robust and can be allowed access to the outdoors at a very young age. To boost the immune system of young poults, place a chunk of sod and dirt from their future pasture into their brooder when they are about a week old. This will help them build early immunity and will minimize the risk of catching a soilborne illness when they do eventually transition fully to the outdoors.

You can give poults limited access to the outdoors by means of a sunporch, which is a screened area built onto an exterior wall of their outdoor brooder or housing. The sunporch should have a wooden floor to keep them off the ground. On sunny warm days, allow poults to spend some time in the sunporch. Be sure to limit their time there if the weather turns cold and if they are not fully feathered.

Temporary Fencing for Poults and Young Turkeys

If you free-range your turkeys, when poults are eight weeks old—old enough to be given full access to the outdoors—consider confining them within temporary fencing for a couple of weeks while they get used to their new surroundings. Poults are very adventurous, and they can easily wander a great distance in a short period of time. You'll want to keep them close to home so they don't wander off and get lost.

Plastic garden fence is easy to install around a poult shelter to keep youngsters safe until they can join the flock.

Another reason to use temporary fencing is to give the new additions time to meet the rest of the flock in a safe space where they can't be attacked by the older flock members as they assert their dominance over the younger ones. It can be dangerous for young poults to be introduced to much larger adult turkeys or other types of poultry too early. A momma hen is fiercely protective of her poults and will guard them from the flock, but in the absence of a momma hen, that responsibility falls on you.

Handling Poults

WHEREAS CHICKS CAN BE SKITTISH and tend to run away when they see a hand coming into the brooder trying to pick them up, poults tend to have the opposite reaction and will usually come right up to investigate. They are generally very easy to pick up from the brooder and will be happy to spend time getting to know you. This is in direct contrast to young chicks, which typically start peeping insistently as soon as you pick them up, demanding to be put back into the brooder.

Of course, there are exceptions to every rule, but early and frequent socializing is the key to calm, friendly turkey poults. Having said this, it's important to keep in mind that poults tend to be more fragile at a young age than chicks, so handle them for just a few minutes at a time to avoid stressing them. Poults can easily become chilled, so if you notice a poult becoming distressed and peeping loudly after spending some time outside the brooder, it's likely cold and should be put back under the heat lamp immediately.

Keep handling to a minimum for the first couple of days so your poults don't become chilled or stressed.

MY FIRST TURKEYS

It didn't take long for my first poults to become my favorite feathered friends on the farm. For three weeks, we raised them in the house under a heat lamp in a large wooden brooder box. This is an especially critical time to make sure the poults are getting enough to eat and drink, their bedding is kept clean and dry, and they don't get chilled. Since I work from home most days, I was able to keep a close eye on them to make sure they got off to a good start.

After three weeks, the poults were getting crowded in the brooder. We began leaving the screened top off the brooder box during the day so that the poults could perch on the top and have a bigger area to hang out in.

For the first few days they behaved themselves and did not venture beyond the edge of the brooder. Then one day while I left the house for a few hours, the poults jumped down onto the floor and began exploring their surroundings. They were soon discovered by my husband and returned to the confines of their brooder, but not before leaving a few presents for me to clean up!

The poults all seemed strong and healthy, so I felt comfortable moving them out to the new turkey coop, which had a heat lamp to keep them warm at night, until their feathers finished growing in. The front portion of the coop was an open-air sunporch, where the poults spent a couple of weeks transitioning from living inside the coop to living outdoors before we eventually gave them access to the pasture.

We kept the turkeys separate from the chickens until they were a few months old, had adjusted to life outdoors, and were at least as big as our head rooster. Then we let the turkeys share the pasture with the chickens.

By the time they were several weeks old the poults had begun to strut and display to sort out the pecking order. These tiny turkeys lowered their wings, spread their tail feathers, and circled each other in a dominance display that was the cutest thing I had ever seen. I couldn't wait to figure out which ones would grow up to be toms and which were hens, but in the meantime, it was lots of fun watching them grow up.

CHAPTER 7

Natural Brooding and Raising

One of the greatest joys of raising poultry is watching a momma hen care for a brood of little ones. Although eggs can be hatched in an incubator, I prefer to let a broody turkey do the work. It avoids the mess of a brooder in the house and the extra chores required to care for the poults in their first weeks of life, plus it is truly heartwarming to observe the interactions between a momma turkey and her poults. With a well-planned setup and a good momma turkey, you can expect up to a 90 percent hatch rate from natural brooding.

There's no sweeter sight than a momma turkey caring for her little ones.

Recognizing a Broody Hen

THE FIRST THING YOU'LL NEED IS A BROODY TURKEY, which is easy to come by because turkey hens regularly go broody. About half of my hens go broody either early in the spring, after laying eggs for several weeks, or a little later in the summer. Some hens will remain broody for a month or so and then return to laying if they are not allowed to keep eggs to hatch, but I've had others that stayed broody for months—no exaggeration.

There's really no mistaking a broody turkey. If you approach her on her nest and reach your hand out to gather her egg, she will puff up and hiss fiercely. If you don't heed this warning, she will peck furiously at your hand or arm, sometimes leaving a bruise or drawing blood if she makes contact. I have taken to wearing gloves and even eye protection when dealing with an especially fierce broody hen (one of them gave me a black eye once). Though sometimes unnerving, the experience can be rather hilarious and will make you think that the saying "Hell hath no fury . . ." should have been written about a broody turkey.

Choosing a Momma Hen

There are relative degrees of fierceness among broody hens. A hen that is interested in interacting with people and accepting of human contact will likely allow you to approach her nest with much less drama than a less friendly hen. As with all poultry, turkeys raised by a momma hen tend to be more wary of people than turkeys hand-raised in a brooder. However, poults that grow up with a calm momma that trusts you will learn to be less skittish around people.

If you have the luxury of choosing a prospective momma from among multiple broody hens, select one that is relatively comfortable around you. It will make things much easier when you need to check on the eggs or to do some quick cleaning. She will also be more likely to trust you to interact with her poults.

Choosing a friendly hen to raise poults will make it easier for you to spend time with the little ones.

Tips for Successful Natural Brooding

SOME YEARS THE HATCHING SEASON GOES SMOOTHLY and successfully with relatively little human intervention, but other years things don't go as planned. The following information should improve your chances for a successful hatch.

Provide a Safe Nesting Area

It's not uncommon for broody hens to start nesting in an area that is unsafe for raising poults, such as out in the relative open where a predator can easily find them. Sometimes they'll squeeze under the coop or try to hide in tight quarters or behind a stack of something where it is difficult to check on them. Turkeys don't need anything fancy, but I strongly recommend providing a private and secure nesting area, with a door that can be closed at night, to keep the soon-to-be momma and her little ones safe. I use a small coop, 4×6 feet, that sits vacant in the turkey yard except when I have a momma raising poults. One or more hens will choose to go broody in this coop every year, which makes my job of keeping the broody hen safe very simple.

This 4 × 6-foot hatching coop provides a safe and convenient nesting site.

Be sure to provide lots of soft nesting material to minimize the chances that eggs will break while they are being incubated. Despite efforts to keep the eggs cushioned in the nest, I find that broody turkeys have an annoying habit of rearranging the nesting material in such a way that the eggs end up lying on the floor with no bedding underneath them. There will usually be some breakage, so give your broody turkey a few more eggs than you are hoping to hatch just in case she breaks a few or if a few aren't fertile.

Separate Broody Hens from the Flock

It can be a challenge to prevent other members of the flock from intruding into the broody hen's nest if it's not hidden away out of sight of the flock. If her nesting location is within easy eyesight of the flock, you may want to consider putting a temporary fence around her. It's especially important to keep tom turkeys away from hens sitting on eggs because eggs can easily get broken during mating. (Yes, toms try to mate with nesting hens, and the hens rarely object.) Worse, a broody hen can be injured by a tom because she will not get up to leave the nest, even if the mating goes on long enough to become dangerous to her. I try to minimize the number of intrusions into the nesting area because that's when eggs get broken. Another downside of a hen nesting in an unsecured area is that I have had chickens get into a nest and eat the eggs.

To prevent intrusions, I put up temporary fencing around the hatching coop as well as a small area of pasture around it to keep curious chickens and other turkeys from going into the nest when the broody turkey is off her nest for her quick bathroom and food breaks.

It's difficult to check on hens that go broody in tight spaces.

Our hatching coop with temporary fence

This method works best if you are around during the day to keep an eye on your broody hen and let her in and out of the fence as needed. When a broody hen gets off the nest, she typically wants to go back to where she's accustomed to eating, drinking, and dust bathing. If she's confined in a different location, she may fly over the fence to go to her usual spots, even if she has access to all necessities inside the fence. If you tend to be a helicopter mom like me, you may find yourself taking on the role of turkey concierge. If there are times when you can't be at home to make sure she gets back inside the fence to her eggs, you can leave the temporary fence around her nest open, but be aware that a few eggs could be damaged if other poultry find the nest when she gets up to take a break.

One Broody Hen per Nest

Turkey hens sometimes like to share nests. The cute sight of two hens on one nest has tempted me into letting them hatch eggs together. While I have had successful hatches with two broody hens sharing a nest, I have also observed broody turkeys competing over eggs, and I've had less than optimal hatches when I have allowed more than one broody turkey in the nest. Now I resist the urge to let them share a nest, and I allow only one broody turkey in the hatching coop.

Don't be surprised if your broody turkey seems to stay on the nest for more than a day without taking a break. Turkeys are very determined brooders, and they can set for a long time between breaks. When she does finally decide to leave her nest for a quick break, the rest of the flock will likely announce her presence in the yard with loud calls and yelps. The broody hen will often be chased around the yard by a hen or two, and you'll hear the fighting purr being directed at her. A turkey doesn't have to be away from her flock for long before the hierarchy gets shuffled.

The broody hen usually takes care of her business and heads back to the nest in about 20 minutes. If your broody hen happens to be nesting near where other hens are laying eggs, check to be certain that your broody went back to sitting on the correct nest. You would think this wouldn't be a problem, but strangely enough sometimes broody hens will be happy to sit on anything, even the wrong nest.

Timing Is Everything

It takes 28 days of incubation for turkey eggs to hatch. It's best if the eggs hatch as close to the same time as possible. Poults that hatch early will be kept waiting under momma hen while she continues incubating the remaining eggs. Eggs that don't hatch within a couple of days of the first to do so will likely be abandoned as momma hen moves off the nest to begin teaching her little ones how to eat and drink.

Once you have a broody hen and are preparing to let her start incubating, it is a good idea to collect eggs daily from all the hens that you intend to hatch eggs from. Keep them safely in an area that maintains a temperature of approximately 55°F (13°C) until you are ready to put them all under the hen at once. Store them pointed ends down in a clean egg carton, and "turn" them once a day (as the hen would do) to keep the yolk properly suspended. To do this, place something a couple of inches high (another egg carton works well) under one end of the carton containing the eggs. Leave it like this for 24 hours, then elevate the other end of the

Don't worry if you occasionally see eggs that aren't under your broody turkey. As she rotates them, she will tuck them back under her.

carton for 24 hours. The best hatching rate is achieved with eggs stored for no longer than seven days. Older eggs can be incubated, but hatchability decreases.

A broody turkey can cover quite a lot of eggs. I've given hens up to 16 eggs, although I've heard stories of wild turkeys incubating up to two dozen. When you have collected enough eggs to hatch, and you are certain that your desired momma hen is broody, give her all the eggs at the same time. If any eggs break early during incubation, resist the urge to add a few more to the nest. Eggs added later won't develop in time to hatch with their siblings, and the presence of unhatched eggs may make the momma hen continue setting on the nest when she should be tending to her poults.

Getting Ready for Hatch Day

IT'S IMPORTANT TO LEAVE THE NEST ALONE as much as possible around the target hatch date and for the first couple of days afterward. Place the feeder and waterer for the poults close to the nest at least a couple of weeks before the hatching date. The last thing you want to do is to introduce anything new or suspicious into the nesting area during those first few days after hatching when the poults are very fragile. You also don't want to trigger momma hen's protective instincts and potentially risk her stepping on her poults if she decides to go into attack mode. Fill the feeder and the waterer a few days before the hatch date, and then let momma hen have her privacy until the eggs hatch.

Turkey mommas tend to be very protective of their babies.

One way to minimize disturbing the momma hen is to install a camera that livestreams to your phone or computer in a location that gives you a good view of the nesting area. The camera should be installed early in the nesting period; better yet, if your broody hens like to hatch in a regular location, you can just leave the camera up year-round in that spot so that it will be in place when you need it. We have a wireless camera installed in a corner of the hatching coop that allows us to watch the activity inside on our phones.

Turkey mommas tend to be very protective of their babies. Before we had the coop camera, it was challenging to get a look at the poults in the first few days after hatching. Whenever I would open the coop door, the momma would call to her poults, and they would all run and hide under her. A coop camera allows you to watch everyone without disturbing them, and you will be able to observe all kinds of things that you would

The first glimpses of the little ones next to their momma are always exciting.

probably never see otherwise. Having a camera is an unobtrusive way to check on the progress of hatching, and it's so fun to see those first tiny heads popping out from under momma.

The First Few Days

WITHIN A COUPLE OF DAYS AFTER THE EGGS HATCH, the momma hen will start to move off the nest for short periods of time. At that point, you should remove any unhatched eggs or broken eggshells, so she won't continue setting on them instead of tending to the needs of her poults. If there are unhatched eggs that appear to still be developing, or possibly have pipped but momma hen is off the nest, you can bring them into the house and place them in an incubator to finish hatching. The poults can be introduced to the momma hen at night when they are a few days old.

A few times a day, the momma hen will move back and forth between the nest and the feeder and waterer. If you happen to be watching a coop camera at the right time, you may catch quick glimpses of the little ones as they scamper over to momma when she changes her location in the coop.

While the hen is off the nest, try to do a quick head count to make sure all the poults are following her and are eating and drinking. If you see a poult that does not appear to be eating and drinking, you may want to move it to a brooder in the house for some TLC. Give it vitamins in its water and regularly encourage it to eat and drink. After a few days, when the poult is hopefully stronger and more energetic, you should be able to return it to the hen. The special bond that forms when poults imprint on you as their primary caretaker in their first few days will remain—it's how I bonded with a couple of my favorite lap turkeys.

It's not uncommon for poults to die unexpectedly in the first few days or weeks after hatching. It's a heartbreaking experience, and you can't help but wonder what happened and if there was anything you should have done differently. Usually, the answer is that there is nothing you could have done to prevent it. Sometimes poults are born with internal problems that unfortunately seal their fate. The first few weeks of life are when most losses occur.

Occasionally, a young poult can be squashed or suffer from some other accident in the coop. I once found a dead poult that had seemed perfectly healthy, so I reviewed the coop camera footage to see what had happened. When the momma turkey went from a nesting position to sitting straight up with her neck outstretched, the little poult somehow got caught between the feathers on her neck. I could see that the poult was peeping in distress, and the momma was shaking her neck trying to release the poult, but to no avail. Sadly, that perfect little poult suffocated, and there was no sign of what had happened when I found it. I could hardly believe such a freak accident had happened, but it helped to know that I couldn't have prevented it.

When the momma hen begins to move off the nest, take the opportunity to quickly check on the little ones to make sure they are doing well.

After a couple of weeks in the hatching coop, the poults are ready to learn how to forage with momma.

Giving Poults a Healthy Start

PROPER NUTRITION IS PARAMOUNT to giving your turkey poults a good start in life. As discussed in Chapter 6, poults should be given a starter feed with 30 percent protein. It's also a good idea to give them powdered vitamins and minerals in their water for the first week to give them an extra boost. When they are about five days old, place a small piece of sod and dirt from their future pasture in the brooder or hatching coop to start building their immunity to soilborne pathogens. Here are a few other things you can do to give your poults the best chance of success.

Minimize Stress

Young poults have fragile immune systems, and they can be easily chilled or stressed. If the weather is cold or rainy in the first couple of weeks after they've hatched, keep the momma hen and her poults inside during the worst part of the day. Waiting until the warmer, dry part of the day is best for introducing them to the outdoors. The poults will be cautious about going outside when the coop door is opened for the first time. The momma hen will be eager to introduce her poults to the outdoors, and she will repeatedly go out and come back into the coop while making sweet purring sounds to encourage her poults to follow her outside.

Momma hen will show her poults how to find tasty things to eat in the pasture, and she should also sit down from time to time and allow the poults to get under her to stay warm. Watch from a safe distance to make certain that momma is doing her job. Happy poults will peep contentedly, while loud distressed peeping is a sign that someone is lost or cold and needs to be reunited with momma hen.

Don't be surprised if, after they've spent the afternoon outdoors, the momma hen tries to nest in the grass with her poults instead of going back into the coop for the night. Of course, it's preferable if you can herd the family back into the coop where they'll be safe from predators. It will also be warmer in the coop than in a nest on the ground, and if the weather turns cold or rainy overnight, the poults will be much better protected. I won't pretend that it's easy to convince them to go back into the coop at night, but keeping everyone safe and healthy is worth the extra time spent getting them back inside.

To shoo them into the coop, I circle around behind the momma with my arms outstretched and walk slowly toward her. As I approach, she will get up and start walking away, and her poults will follow her. I stay a few feet away from her so that she doesn't go into protective momma mode and try to attack. I talk to the hen in a soothing voice (I pretty

It's best for the hen and her little ones to be safe in the coop at night.

much always talk to my turkeys) and point my arms in the direction of the coop. It may sound strange if you haven't seen it, but I really do believe that when I point toward the coop and say "back to the coop," the momma hen knows what I want her to do.

Finally, avoid chasing or panicking the poults in your attempts to get a snuggle during the early days after hatching. Too much stress can be detrimental to young poults, much more so than it is for chicks. By the time the poults are a couple of weeks old and are running around outside with momma, you can start getting them accustomed to being handled by sitting with them and offering treats such as dried mealworms or grubs.

Offering treats (always in moderation) is an excellent way to bond with your poults.

Provide Protection from the Flock

Although turkeys are much larger than chickens, turkey poults are just as tiny as chicks. It's important to minimize the risk of the little ones getting stepped on. This can easily happen if the momma hen gets excited while chasing away a member of the flock that comes too close for comfort, or if a tom turkey tries to get romantic. To minimize the potential for dangerous situations, it is helpful to separate the momma and little ones from the rest of the flock for a few weeks. I use temporary plastic garden fence around the hatching coop to give momma turkey and her poults a safe area for the little ones until they are strong enough to meet the rest of the flock.

I recommend using a fence with small openings, such as plastic garden fence with 1-inch holes. You'll be surprised how easily a poult can slip through a barrier to get outside its enclosure and then be unable to figure out how to get back in, causing all kinds of distress for momma. It makes it easier on both of you if you can prevent a poult from getting out in the first place. Young poults are excellent escape artists, and you will find yourself having to retrieve them from outside the fence and bring them back in to momma,

A 3-foot-tall plastic garden fence around the hatching coop is a simple temporary way to separate young poults from the flock.

A secure run covered with ½-inch hardware cloth is another option for protecting momma and her little ones from the flock for a few weeks.

but at least in the early days it's helpful to try to keep everyone safely together inside a temporary fence.

Alternatively, if you have a secure run that the rest of the flock can do without for a few weeks, move the momma and poults into that. It must contain a small, enclosed structure to protect them from the weather. If there isn't a coop with a locking door inside the run, then the walls of the run should be built from ½-inch hardware cloth so that it is predator proof.

TIPS FOR SUCCESSFUL NATURAL BROODING

- Provide a safe nesting area.
- Allow just one hen per nest.
- Minimize stress for the momma and her poults.
- Provide proper nutrition for the poults.
- Provide protection from the flock.

FIRST DAY OUT

Spring is always an exciting time on the farm because it is the season for hatching turkeys. My momma hen Eleanor was quite attentive to her poults, being very careful where she put her large feet when walking around in the coop so as not to step on the little ones. However, she would occasionally step on the babies if she got a bit excited because I came too close to them, and that's when I knew it was time for me to close the coop door and let them have some alone time. For the first week I did not see much of her poults since it was relatively cold outside, and they stayed under Eleanor much of the time except for short periods of eating and drinking.

At about a week old, the little ones started spending more time out from under their mom. Eleanor was still very protective of them, and every time I would open the coop door to refill the feeder, she would make an alarm call and the babies would go dashing back under her for safety. When the poults reached 10 days old, the weather warmed up enough that Eleanor brought her poults outside the coop for the first time.

Once the poults were several weeks old, Eleanor was slightly less protective, and they spent more time out from under her. I could tell when Eleanor was ready to have a larger area of pasture to roam because she started to fly over the fence around the hatching coop. Luckily, she did not go far, and it was easy to shoo her back in with her babies. The little ones were already jumping up onto a roost a foot off the ground, so I knew they would be flying out of their pen to escape into the bigger pasture along with their mom in no time!

I had a feeling the poults' great escape would probably be right around the time that I left for my first vacation in quite a while. Every time I spend a night away from the farm it seems like all manner of chaos breaks loose, but when you are lucky enough to live the farm life, surrounded by charismatic birds like turkeys, you find that you really don't want to leave all that often!

Left: *A protective momma keeps her poults away from the rest of the flock until they can fend for themselves.*
Right: *Poults are excellent flyers and are able to roost with their momma after just a few weeks.*

Introducing Poults to the Rest of the Flock

The momma turkey will let you know when she is ready to rejoin the flock. She will start spending more time outside with her poults, and you'll notice them sneaking under or flying over their fenced enclosure to go on adventures with greater frequency. At night, instead of sleeping on the nest, momma hen will begin roosting again. Poults are excellent flyers from a young age, and by the time they are a few weeks old they are ready to follow her. When everyone starts sleeping on the roof of the hatching coop, I know they are ready to join the rest of the flock.

After I take down the fencing, momma hen remains very protective of her poults for several weeks after rejoining the flock. It's common to see her and her babies in the outskirts of the yard, where they will be less noticed by the other turkeys.

At night the momma hen will roost with the flock, with her poults sleeping under her wings until they get so large that only a few can fit, although she will try her best to cover them all. This is one of my favorite sights—seeing the poults snuggled up close to momma and her covering the poults with her huge fully outstretched wings on the roost, trying to protect them from their new and very pecky roost mates! Often a momma hen will settle onto the roost with her little ones earlier than the rest of the flock. As the other turkeys fly up to roost, the more dominant members of the flock may harass the momma until she and her poults fly down to the ground or to a lower roost. Eventually momma will find a spot on the roost, away from the most troublesome members of the flock, where she and her poults can roost undisturbed for the night.

Determining the Sex of Young Turkeys

BY THE TIME POULTS ARE ABOUT SIX WEEKS OLD, it's possible to make educated guesses about their sex. For poults hatched in the same brood, the young males will have slightly longer and thicker legs, larger feet, and larger bodies than the young females. Another method is to look at the size and shape of their snood. The snood is apparent from an early age in males, appearing as a small pink bump above their beak. Young females also have a small snood, but it is usually less noticeable than the snood on young males. The snood size difference is subtle in young turkeys but becomes more noticeable as they age.

As the snood grows longer on juvenile males (known as jakes) it will begin to fall to one side of the beak. When you see a snood hanging to the side of the beak, you can be sure of the sex of male turkeys. The snood of the juvenile females (known as jennies) also grows longer with age, but it typically reaches a maximum length of about a centimeter and will either sit semi-upright or lie flat on top of the beak.

The shape of the snood is also an indicator of sex. The male snood is wider at the base, where it is attached to the beak, and narrower at the tip. When the male snood is retracted, it points upward in a triangular shape, resembling a little unicorn horn. The female snood is narrower at the base, and the width is the same along its entire length. Another indicator of the sex of turkeys is the size of the caruncles, which are the bulbous protuberances on their throat. By the time males reach a few months old, their caruncles are noticeably round and red in color, while in females the caruncles are smaller and tend to be pink in color.

At six weeks old, the size difference between the snoods of males and females starts to become discernible.

CHAPTER 8

Healthcare

Many people who keep backyard poultry consider their birds as pets and go to great lengths to ensure their well-being. Others view them as livestock, and while they still want the best for them, they have a less emotional approach to providing healthcare, including monetary or other considerations. We each must do what we feel is best given our situation and resources. Whatever your viewpoint, it makes sense to establish a relationship with a veterinarian who treats farm animals or has some experience with avian patients.

Proper nutrition, attention to clean indoor and outdoor living conditions, limiting contact with wildlife that can be vectors for disease, and providing well-ventilated shelters are important to ensure that turkeys have a good start in life and remain healthy. If you follow these best practices, you will find that heritage turkeys generally remain free from health issues. Broad-breasted varieties differ from heritage turkeys in that they are prone to weight-related ailments as they reach increasingly large size at maturity.

Even with the best of care, occasionally illnesses occur, but I want to reiterate that you may not experience any of the conditions described below if you have heritage turkeys. Here are a few of the more common health issues and diseases to be aware of, just in case.

Diseases

DISEASES CAN BE SPREAD TO TURKEYS through direct contact with wild birds and domestic poultry or their droppings, by insects, from water sources used by wild birds, and via soilborne organisms. Good sanitation practices and biosecurity measures are important for ensuring the health of your turkeys. Several of the following conditions have limited or no treatment options, so prevention is the only answer.

Avian Influenza

Avian influenza is a virus spread from wild waterfowl to domesticated poultry through droppings or contact with infected individuals. A common source of transmission is when wild and domesticated birds gather at communal water sources such as streams or ponds. There are both low-pathogenicity and high-pathogenicity forms of the virus. The low-pathogenicity form causes little or no signs of illness, but the high-pathogenicity form causes high mortality and can spread rapidly among wild populations and to domesticated poultry flocks.

The USDA began vaccine trials for highly pathogenic avian influenza (HPAI) in 2023. At the time of this writing, there is no vaccine available for backyard flocks and no treatment. It is standard practice to cull birds that are infected with HPAI to avoid further spread. A large outbreak of HPAI from 2022 to 2023 resulted in the culling of nearly 78 million birds in commercial and backyard flocks across the United States. Symptoms of avian influenza include the following.

- Sudden death without clinical signs
- Lack of energy, movement, or appetite
- Swelling of head, comb, eyelid, wattles, and hocks
- Purple discoloration of wattles, comb, and legs
- Nasal discharge, coughing, and sneezing
- Nervous system signs, tremors, or lack of coordination
- Diarrhea

Birds affected with the highly pathogenic form typically die within three days of displaying one or more of these symptoms. If you experience sudden unexplained deaths in your flock or notice these symptoms, contact your local state wildlife agency to report it. Staff may want to collect the birds for testing, and they will coordinate with other wildlife management agencies as needed and assist with euthanizing affected flocks. Obviously, this is a nightmare scenario that none of us want to experience. It is best to avoid the potential for contracting this disease by keeping wild waterfowl away from your flock and not allowing turkeys access to areas frequented by waterbirds. Access to natural water sources may be difficult to control for truly free-range turkeys, but you can minimize the risk to pastured birds by fencing them away from natural water sources.

Veterinary Care

It's only natural to want to provide the best care for our birds when they become ill. Sometimes a health condition is serious enough to warrant professional medical care, but it can be difficult to find a veterinarian with experience treating domestic poultry. Before you find yourself in need of a veterinarian, it's a good idea to contact the veterinarians in your area to find out if there are any that specialize in treating poultry and would be willing to see your turkeys.

Although there are increasing numbers of veterinarians willing to accept poultry patients, many avian veterinarians primarily treat exotic pet birds and may not be as experienced treating domestic poultry. When searching for an avian vet, try to find one who specializes in poultry, or better yet, one who is also a poultry owner. If you can't find a veterinarian in your area, your state veterinary college or extension agent may be a resource for finding veterinary care or obtaining laboratory diagnosis.

The field of medicine for domestic poultry is much less advanced than it is for other types of animals. Despite the best efforts of the veterinarian, there is just not as much information available regarding diagnosis and treatment in domestic poultry. I say this based on many years of experience with chickens and turkeys and more veterinary visits than I can count. Sometimes you get lucky and will have a win when taking poultry to the vet for treatment, but most often when a member of the flock is ill enough to warrant medical attention, the outcome is not what we hope for.

Blackhead Disease

One of the most serious and more common ailments discussed in the backyard turkey community is blackhead disease, also known as histomoniasis. This disease is caused by a protozoan parasite (*Histomonas meleagridis*). Blackhead disease can be contracted in a variety of ways, including foraging in areas where infected bird droppings are present or eating infected invertebrates such as earthworms. Roundworm (*Heterakis gallinarum*), a parasite often present in the intestines of chickens as well as other poultry and wild birds, is frequently involved as an intermediate host. The name "blackhead" is misleading because birds affected by this disease rarely display a discolored head. Bright yellow sulfur-colored diarrhea in conjunction with listless behavior and loss of appetite are the telltale signs of this disease. According to the *Merck Veterinary Manual*, mortality in turkeys is 80 to 100 percent.

Check with your local agricultural extension office to find out if blackhead disease is present in your area. If it is, do not keep your turkeys with chickens. As of this writing, there are no approved treatments or vaccines for blackhead. Anecdotally, some people have had success treating early cases by adding ground cayenne pepper to their turkeys' feed at the first sign of symptoms.

Bumblefoot

Bumblefoot is caused by the *Staphylococcus* bacteria and occurs on the bottom of the foot or sometimes between the toes. It appears as a small brown or black scab, under which typically is a pus-filled abscess. Swelling of the skin around the scab occurs to varying degrees, sometimes making a round protrusion around the center of the infection. As the severity of the infection worsens, it can cause limping or difficulty walking. If left untreated, the wound develops necrotic inflammation, resulting in dead tissue.

Early bumblefoot can be treated by soaking the affected foot for 10 minutes or so in an Epsom salt bath to soften the skin and then gently prying up the scab and removing the underlying abscess. Repeated soaking and gently prying out the scab and abscess with either your fingernail or a dull implement is preferable to a more invasive surgical removal because the healing time will be reduced. Afterward, disinfect the wound, apply a topical antibiotic, and wrap the wound to keep it clean for a few days until it heals. Note that this procedure is generally possible only in tame birds that don't mind being handled. It is easier to have two people, one to hold and calm the bird, and one to treat the foot.

Coccidiosis

Coccidiosis is caused by a protozoan parasite in the genus *Eimeria*. It is primarily a disease of poults within the first few weeks of life, but it can occur up to eight weeks of age. Natural immunity typically develops as birds mature. Wet bedding is a primary factor in allowing this disease to flourish, and regular attention to providing a dry and clean brooder environment can prevent it. Affected birds are noticeably less active, appear hunched, and stop eating. A telltale symptom is the presence of blood in the droppings. Mortality is high if the condition is not treated at the first sign of symptoms. The most common treatment is amprolium, which is administered in the water so that all birds that are possibly affected can be treated.

Fowl pox will clear on its own, but it may be helpful to provide daily cleaning of the lesions and supportive care.

Fowl Pox

Fowl pox is a viral disease caused by the fowl pox virus, which is primarily spread by mosquitoes. There are two forms of the disease, a dry form and a wet form. The dry form results in small, round lesions on the unfeathered areas of the head and neck. The lesions can start out as pale, raised bumps before gradually turning dark and forming scabs that will fall off in a few weeks. The darkened lesions of the dry form can be similar in appearance to pecking or fighting injuries. In cases where lesions develop around the eyes, swelling and discharge can cause the eyes to seal shut. Try to keep the eyes clean with a daily flush of saline solution.

The wet form is more serious and causes lesions in the mouth and throat that can make it difficult for affected birds to eat and can cause respiratory distress. There is no treatment for fowl pox once it is acquired. Both forms of the virus will run their course and clear on their own, but in more serious cases, supportive care such as isolating the affected bird, minimizing stress, and providing extra nutrition can be beneficial.

Try to eliminate the potential for this virus by limiting sources of stagnant water that provide mosquito habitat. In the summer, change water in troughs and tubs frequently to prevent them from becoming mosquito hatcheries.

Respiratory Infections

The independent nature of turkeys and their dislike of enclosed shelters even during adverse weather conditions can put them at risk for respiratory infections. Providing them with a variety of open-sided and high-roofed structures that they will be more inclined to use during cold weather minimizes that risk.

Proper ventilation is important in preventing respiratory illnesses.

Another important way to minimize the potential for respiratory disease is to provide proper ventilation in enclosed nighttime living quarters. Moist bedding and humid conditions allow for the buildup of ammonia, which is harmful for a turkey's respiratory system and can also trigger fungal respiratory infections. Good ventilation is especially important in the fall and winter, when turkeys are likely to spend more time in shelters kicking up dust. Turkeys, especially broad-breasted varieties, can easily suffer from nasal congestion due to dust getting in their nares.

Respiratory disease can occur due to a variety of viral, bacterial, and fungal agents. Symptoms include discharge from the eyes or nose, bubbling in the corner of the eye, congestion, swelling of the face or wattle, sneezing, coughing, gasping, rales, or shaking of the head. Some respiratory infections are mild and may clear on their own; others can spread quickly through the flock, causing mortality in multiple birds. Determining the cause of infection can be difficult because the symptoms are similar among a variety of respiratory illnesses.

If you suspect a respiratory disease, it is always best to get a veterinary diagnosis first instead of administering over-the-counter drugs without knowing what exactly you are trying to cure. Treating an unknown condition with a drug that may not be effective against it doesn't do your flock any good, or anyone else's flock for that matter, as it only serves to contribute to broader drug resistance among microorganisms.

AVIAN MYCOPLASMOSIS is one of the more common respiratory infections in poultry. It is caused by several species of the *Mycoplasma* bacteria. Common symptoms include watery eyes, swelling of the sinuses, respiratory distress, listlessness, reduced feed intake, and weight loss. A poultry veterinarian can use a tracheal swab or a blood sample to test for mycoplasma. Mycoplasma infection can be treated in some cases with antibiotics available over the counter or from a veterinarian. It can also be fatal if complicating factors are present or in individuals with secondary infections. It is not uncommon for backyard flocks to suffer losses of multiple birds due to a mycoplasma infection. A bird that recovers from mycoplasma will be a carrier for life and can potentially spread the bacteria to other birds, even if no outward symptoms are present.

Parasites

PARASITIC INFESTATIONS are generally less serious than the diseases just described, but they can lead to losses if not recognized and treated. A variety of mites and lice infect turkeys. These external parasites are usually kept in check through behaviors including dust bathing, sunbathing, and preening. However, there are occasions when environmental factors may limit these behaviors and infestations can occur. This is particularly common in the winter, when poor weather can reduce the frequency of dust bathing. External parasites can also get a foothold if a bird becomes ill and fails to keep up with grooming itself.

Body Mites and Lice

The vent is the most obvious location where external parasites can be spotted. It is good practice to occasionally check vents for the presence of mites, lice, or their egg sacs, which are found at the base of the feather shafts. For individuals that dislike being handled, inspecting them at night when they are roosting and tend to be more docile is the best method.

For mild cases, when only a few lice or mites are seen, food-grade diatomaceous earth (DE) can be sprinkled on the skin and at the base of the feather shafts where parasites are present, and DE or wood ash can also be added to the dust bath as a preventive measure. Take care not to breathe in the very fine particles as they can cause respiratory irritation. For more serious infestations, use a stronger pesticide such as a poultry dust containing permethrin or pyrethrum or a spray made with Elector PSP. Turkey housing should also be cleaned and pesticides applied to roosting bars and in cracks and crevices, where mites and lice like to hide.

Scaly Leg Mites

These mites occur under the scales on the legs and feet. This condition is easily spotted on infected birds because the scales on their legs become raised and roughened rather than lying flat and smooth. If left untreated, this condition can lead to deformation and difficulty walking or even lameness or necrosis in serious cases. Treatment is achieved by smothering the mites. Home remedies to treat scaly leg mites include petroleum jelly, vegetable oil, or VetRx. Slather well on the legs and feet and repeat applications every few days for a couple of weeks to achieve good results. Do not dip the legs in gasoline or kerosene, even if you see this old-school method recommended online!

Intestinal Worms

Several species of intestinal worms can infect turkeys. Roundworms are one of the more common culprits, and the large roundworm can be seen by the naked eye in the droppings as a 2- to 3-inch-long worm about the width of a pencil lead. Although it can be alarming to see worms in the droppings, the presence of a small number of worms does not generally lead to any serious health problems. If, however, an infected bird is showing symptoms such as diarrhea, depression, weight loss, or reduced egg production, it may be advisable to have a fecal float test done at the veterinarian's office to determine the parasite load and the appropriate course of treatment. Wormers are specific to worm species, and only a limited number of wormers are approved for use in turkeys.

Health Conditions of Broad-Breasted Turkeys

BROAD-BREASTED TURKEYS ARE PRONE to developing several health conditions with increasing age and body size. Some of the more common issues that occur in very large turkeys include lameness, joint issues, broken bones, and heart or respiratory failure. The potential for these

Broad-breasted turkeys kept past maturity often experience weight-related health conditions.

problems to occur can be minimized with attention to their diet and by limiting feed intake.

Broad-breasted turkeys tend to overeat when given free access to feed. It's best to manage feed intake by providing only the amount they need once or twice a day, and to provide access to pasture to encourage foraging for some of their daily intake, as well as exercise. Limiting food intake and encouraging active behavior can help keep broad-breasted turkeys at a healthier weight and extend their lifetime. The potential for leg and feet injuries can also be minimized by managing their weight and by providing low roosts or hay bales for mature males, to discourage them from roosting up high and incurring injuries when flying down.

Difficulty walking or inability to walk, increasingly sedentary behavior, and labored breathing are signs of declining health in broad-breasted turkeys. It is always hard to watch one of our animals suffer from decreased quality of life, and it is a deeply personal decision whether to euthanize an animal under one's care. No one knows our animals as well as we do, and only you can decide whether an individual's quality of life is diminished enough to warrant this decision.

Turkey Hen Health

MATING INJURIES AND REPRODUCTIVE DISORDERS can occur in turkey hens, though they are not common in my experience. Mature toms of both heritage and broad breasted varieties can injure hens with their spurs during mating. Pulled-out or broken feathers and small patches of bare skin on a hen's back are not usually too serious, but the spurs can cause deep cuts under a hen's wing, sometimes down to muscle.

In the case of such an injury, clean the wound and isolate the hen until it heals. Turkeys are amazing at healing, even from deep wounds. Just keep the wound clean and provide supportive care and nutrition. You may want to add vitamins and electrolytes to her water.

Trimming the sharp tips of a male's spurs with a pet nail clipper can minimize the potential for mating injuries. Don't trim too much or you may expose the "quick" (blood vessel), causing the spur to bleed. A good time to trim spurs is at night when the male is calmly sitting on a roost.

Another issue is that a particularly heavy tom can injure the hens during mating, especially if the flock is confined in close quarters and the mating becomes excessive. A large tom may need to be separated from the hens if this becomes an issue.

Reproductive Disorders

Although reproductive issues do occur in turkeys, they are much more likely to happen with chickens, which have been developed to maximize egg production. When disorders occur in turkeys, they are typically serious. Issues include egg binding, internal laying, egg yolk peritonitis, and vent prolapse. These are all serious conditions that are often fatal without prompt diagnosis and treatment—and sometimes they are fatal even with treatment.

Fortunately, reproductive disorders in turkeys are not very common.

Symptoms of reproductive disorders include the hen choosing to self-isolate, becoming hunched and inactive, and having less interest in feeding. The hen may have a swollen abdomen, depending upon how long she has been hiding her symptoms, and often a messy discharge from the vent. Isolate the hen in a safe area so that she won't be pecked by other members of the flock while you attempt to make a diagnosis and evaluate your treatment options.

EGG BINDING is when an egg gets stuck in the reproductive tract, and the hen is unable to pass it. The hen can in some cases be assisted with warm, moist heat, such as an Epsom salt bath, or by gently lubricating the vent while taking care not to break the egg inside. Depending on the size of the egg or the severity of the egg binding, it may not be possible for you to assist the hen in passing the egg. Treatment also depends upon being able to handle your bird without causing unnecessary stress, and this may be no easy task. Veterinary intervention and surgery may be an option if you can find an avian veterinarian willing to do it.

INTERNAL LAYING/EGG YOLK PERITONITIS occurs when a yolk gets off track, and instead of traveling from the ovary to the oviduct, it gets deposited in the abdomen. Egg yolk peritonitis is the condition that occurs when the yolk material causes inflammation of the peritoneum, which is the membrane that lines the abdominal cavity. The inflammatory reaction results in the accumulation of fluid in the abdomen. In addition, the yolk material can become infected with bacteria (most commonly *E. coli*) that can then spread into the bloodstream, causing infection in other organs. If it is diagnosed early, a veterinarian can treat this condition by draining the abdominal fluid, prescribing antibiotics, and potentially inserting a hormone implant to stop egg laying, which will give the hen time to recover. But even with these interventions, in serious cases the infection may be too advanced to be successfully treated, in which case the condition is fatal.

VENT PROLAPSE occurs when the lower portion of a hen's oviduct everts to lay an egg, but instead of retracting as it normally does it stays outside the body. This condition is visible as tissue protruding from the vent. The prolapse can sometimes be treated in less serious cases if it is caught early. Clean the area carefully and press the tissue gently back inside the vent with a gloved finger. Use either Preparation H or raw honey to treat the area a couple of times a day. Keep the hen isolated and in dim light (to try to delay the egg-laying cycle) to allow her time to recover. More serious cases where the vent cannot be pushed back in, or if additional injuries have occurred due to pecking by the flock, may not be able to be successfully treated. Vent prolapse is more likely to occur in broad-breasted than in heritage turkeys.

First Aid

It's stressful enough having a health emergency in your flock, and it's even worse when you have to run around the house or dash out to the farm store to gather the first-aid supplies you need. Here are a few things I keep on hand in a designated first-aid box for my poultry.

FIRST-AID KIT

- Blood clotting powder
- Blu-Kote antiseptic
- Cotton balls or swabs
- Disposable gloves
- Epsom salt
- Gauze
- Iodine or Betadine
- Pet nail clippers
- Petroleum jelly
- Preparation H
- Saline solution
- Small scissors
- Small syringes
- Topical antibiotic
- Vet wrap
- Vetericyn Plus Antimicrobial Poultry Care spray
- VetRx Poultry Remedy
- Vitamins and electrolytes

Glossary

BEARD. A cluster of coarse, bristlelike modified feathers that hang down from the breast of a male turkey and some female turkeys.

BROODER. A safe, enclosed, heated space for raising young poultry when they aren't raised by a mother hen.

BROODY. Nesting behavior that occurs when a hen wants to hatch eggs.

CARUNCLE. Fleshy bumps on the base of a turkey's throat and on the head and neck.

DRESSED. A turkey carcass that is prepared for cooking (plucked and organs removed).

FEED CONVERSION RATIO. A measure of an animal's efficiency in converting feed into body weight.

FLOGGING. An act of aggression whereby a turkey attacks with either his feet or his wings.

GRIT. Small pebbles turkeys ingest and use to grind food in the gizzard.

HARDWARE CLOTH. A woven wire mesh that provides better protection from predators than chicken wire.

HEN. An adult female turkey.

JAKE. A juvenile male turkey.

JENNY. A juvenile female turkey.

NATIONAL POULTRY IMPROVEMENT PLAN (NPIP). A voluntary state and federal cooperative testing program that certifies the health of industrial and hobby flocks.

POULT. A young turkey.

RAFTER. A group of turkeys.

SNOOD. Fleshy protuberance on top of the beak of a turkey.

SPRADDLE LEG. An injury seen in young poults resulting from the legs slipping apart into a spraddle or splayed-leg position.

SPUR. A sharp, pointed protrusion on the shank of the leg in male turkeys that is used for fighting.

STANDARD OF PERFECTION. The physical appearance and coloring traits that define the breed standard for poultry.

STRAIGHT-RUN. Unsexed (poults that have not been sorted by sex).

STRUTTING. A display by a tom turkey with his tail fanned out and wings held down and away from his body.

TOM. An adult male turkey, also known as a gobbler.

TREADING. Mating behavior that occurs when a male turkey appears to walk in place on the back of a hen.

TURKEY MATH. The habit of continually wanting to expand the size of your turkey flock.

VENT. The exterior opening for the digestive, urinary, and reproductive tracts in poultry, used to expel feces and lay eggs. Also called the cloaca, it is found under the base of the tail.

WATTLE. A thin flap of loose skin that hangs down under the chin of both male and female turkeys.

Resources

Books

Butchering Poultry, Rabbit, Lamb, Goat, and Pork by Adam Danforth, Storey Publishing, 2014

Hatching & Brooding Your Own Chicks by Gail Damerow, Storey Publishing, 2013

Storey's Guide to Raising Turkeys, 3rd Edition by Don Schrider, Storey Publishing, 2013

Organizations and Government Agencies

The Livestock Conservancy
https://livestockconservancy.org

US Centers for Disease Control avian influenza information
https://cdc.gov/flu/avianflu/index.htm

US Department of Agriculture directory of local agricultural extension service offices
https://nifa.usda.gov/land-grant-colleges-and-universities-partner-website-directory

US Department of Agriculture Poultry Extension
https://poultry.extension.org

A list of respiratory diseases and their symptoms can be found here:
https://poultry.extension.org/articles/poultry-health/diseases-of-the-poultry-respiratory-system

Online resources

Backyard Poultry magazine
https://backyardpoultry.iamcountryside.com

GRIT magazine
https://grit.com

Merck Veterinary Manual
https://merckvetmanual.com

Modern Farmer
https://modernfarmer.com

Mother Earth News
https://motherearthnews.com

PoultryDVM
https://poultrydvm.com

Hatcheries Selling Heritage Turkeys

Cackle Hatchery
https://cacklehatchery.com

Hoover's Hatchery
https://hoovershatchery.com

Ideal Poultry
https://idealpoultry.com

Meyer Hatchery
https://meyerhatchery.com

Murray McMurray Hatchery
https://mcmurrayhatchery.com

Porter's Rare Heritage Turkeys
https://porterturkeys.com

Privett Hatchery
https://privetthatchery.com

Stromberg's
https://strombergschickens.com

Welp Hatchery
https://welphatchery.com

Acknowledgments

Many thanks to Karen, David, and Anna, who provided me with comments as I was writing this manuscript and who shared their experiences with me to help make this book well-rounded and representative of life with turkeys from the West Coast to the East Coast, whether keeping heritage or broad breasted, with a variety of housing and pasture setups. I also want to thank the turkey community on Instagram for sharing the love of turkeys with me and for the many chats and smiles we've shared through the years over the antics of these charismatic birds.

Index

Page numbers in *italics* indicate photos; numbers in **bold** indicate charts.

A

American Poultry Association (APA), 21, 23
anatomy, turkey, 44, *44*
avian influenza, 134–35
avian mycoplasmosis, 138–39

B

beard of tom turkey, 44, *44*, 45, *45*
bedding material, brooder and, 99–100, *100*
behavior, 25–26, *25*, *26*, 43. *See also specific behavior*
blackhead disease (histomoniasis), 26, 27, 136
body mites, lice and, 139
bonding, treats and, 127
broad breasted turkeys, *15*, 16
  artificial insemination for, 19
  bronze varieties, 18, 19
  choosing to raise, 17, 25
  color varieties, 19
  factory farming, 15, 17
  health conditions/issues, 19, 140–41, *140*
  heritage turkeys compared to, 16
  meat, raising for, 37–38
  raising, 18–19
  Thanksgiving and, 17, 18, 19, 66, 82
  wild turkeys and, 14
  roosting needs for, 72
Bronze varieties, 18, *18*
  broad-breasted turkeys, 19
  heritage, 23, *23*
  "Standard Bronze," 18, *18*, 21, 23, *23*, 25
brooder, *97*
  bedding material, 99–100, *100*
  checklist for, 98
  chicks added to, 106
  cleaning, 103
  heating options, 101–3, *101*
  indoor, toxic fumes and, 97
  poults outgrowing, 110
  setting up, 97–99
  size and space per poult, 98–99
  temperature, monitoring, 102
brooder plate, 103
brooding, natural. *See* natural brooding/raising
broody hen(s), *121*
  one per nest, 120–21
  recognizing, 116–17
  separating from flock, 119–120, *119*
bumblefoot, 136–37

C

calcium, 84–85
calls/sounds, 58–61
caruncles, *44*, 45, *45*
chickens, turkeys living with, 26–31
  benefits of, 29
  chicks in brooder, 106
  coexistence and, 27–28, *27*, *28*
  risks of, 29
  turkey tribe story, 30–31, *30*, *31*
chickens, raising, 10
chicks, brooding versus poults, 94, *94*
clipping wings, 77, *77*
cloaca (vent), 46
coccidiosis, 137
conservation, importance of, 24–25
coop(s)
  determining need for, 66–67
  failure/lessons learned, 68
  hatching, 118, *118*, *120*, 127, *127*
courtship, mating and, 61–62, *61*, *62*
critical breeds, 24

D

diseases, 134–39
dominance
  flock dynamics and, 53–55
  poults and, 48
  tom turkeys and, 53, *54*, 55
  you are the boss, 55–57, *56*, *57*
dust bathing, 87, *87*

E

egg binding, 142
egg yolk peritonitis, 142
eggs, turkey, 11, 34–37
  incubation of, 121–22, *121*
  eggshells of, 20, *20*, 35, *35*
  cooking with, 36
  nesting places and, 35–36, *36*
  selling, 34
  size of, compared to chicken, 37, **37**
electric fencing, 74–75, *75*

F

facial colors, 47, *47*
factory farming, 15, 17
feed. *See* nutrition; nutrition/feeding, poults
feed store, poults and, 95
feeders, 82, *82*, 104–6
  hanging, *104*
fencing
  electric, 74–75, *75*
  hatching coop and, 127, *127*
  temporary, 111–12, *111*, *120*
"fighting purr," 59
first-aid kit, poultry, 143
first turkeys story, 113
flock
  introducing poults to rest of, 130–31
  protecting poults from, 127–28, *127*, 130, *130*
  tom-to-hen ratio, 54–55
flock dynamics, 53–55
  human caretakers and, 55–57, *56*, *57*
fowl pox, 137
free rangers, predators and, 74–77
fun with turkeys, 38–39, *38*, *39*

G

games, playing, 39, *39*
gobble, tom turkey, 58
grit, 84, *85*
  starter feed and, 108, *108*

H

handling turkeys, 25–26
  poults, 26, *26*
hatcheries, 95–96, *96*
hatching coop, 118, *118*, *120*, 127, *127*
healthcare, 133–143
  broad-breasted turkeys, 140–41, *140*
  diseases and, 134–39
  first-aid kit, poultry, 143
  hen health, turkey, 141–43
  parasites and, 139–140
  veterinary care, 135
heat lamp, 101–2, *101*
  correct bulb for, 102–3
heating options, brooder, 101–3
  brooder plate, 103
  heat lamp, 101–2, *101*
hen(s). *See also* broody hen(s)
  calls of, 60
  tom-to-hen ratio, 54–55
hen health, turkey, 141–43
  mating injuries, 141
  reproductive disorders, 142–43
herbs, probiotic supplements and, 109
heritage turkeys, 14–15, *16*, *21*
  broad-breasted turkeys compared to, 16, *16*
  bronze varieties, 18, 23, *23*
  characteristics/origin of, 19, 22–23
  choosing to raise, 17, 25
  genetic diversity and, 24
  meat, raising for, 38
  raising, 19–21
  toms, weight at maturity, 37
history, domestic turkeys, 14–17
housing needs, poults, 110–12
  brooder and, 110
  temporary fencing, 111–12, *111*
  transitioning poults outside, 110–11

I

incubation of eggs, 121–22, *121*
influenza, avian, 134–35
internal laying, hens and, 142
intestinal worms, 140

L

"lap turkey" story, 40-41, *40*, *41*
leg mites, scaly, 139–140
lice, body mites and, 139
Livestock Conservancy, The, 19, 24–25

M

mail-order poults, 96, *96*
mating
  artificial insemination vs., 19
  courtship and, 61-62, *61*, *62*
mating injuries, hens, 141
meat, raising turkeys for, 37–38. *See also* Thanksgiving turkeys
mites
  lice and, body, 139
  scaly leg, 139–140
mycoplasmosis, avian, 138–39

N

Narragansett breed, 25
  description, 22, *22*
natural brooding/raising, 115-131, *116*. *See also* broody hen(s)
  first day out story, 129
  first few days, 123–24, *124*
  hatch day and, 122–23, *123*
  hatching coop, 118, *118*
  momma hen, choosing, 117, *117*
  poults, healthy start for, 125–28, *126*, *127*, *128*
  poults, introduction to flock, 130–131
  timing is everything, 121–22
  tips for success, 118–122, 128
nesting places, 35–36, *36*
nutrition, 80–82
  developer/grower feed, 80, *81*
  feed efficiency and, 81–82
  layer/maintenance feed, 81, *81*
  treats, 84, *84*
nutrition/feeding, poults
  feeding chart for, **108**
  natural brooding and, 125
  starter and grower feed, 80, *81*, 108
  supplements, *108*, 109
  treats, 110, 127

O

outdoor space, 73

P

parasites, 139–140
  dust bathing and, 87, *87*
pasture rotation, 85–86, *86*
poults (young turkeys), 93, *99*, 108–10. *See also* brooder; flock; nutrition/feeding, poults
  brooding, versus chicks, 94, *94*
  color varieties and, 19
  coop for, 66
  feeding time and, 80, *81*
  first experience raising, 113
  flight/roosting behavior, 130, *130*
  handling, 112, *112*
  housing needs, 110–12
  natural brooding/raising, 125–28, *126*, *127*, *128*
  purchasing, 95–96, *96*
  sex of, determining, 131, *131*
  strutting behavior of, 48
  temperament, behavior and, 25–26, *25*, *26*
predators, 74–76
  electric fencing and, 74–75, *75*
  free birds/escapees and, 75–76, *76*
  wing clipping and, 77, *77*
  winter and, 90
probiotic supplements, herbs and, 109
purring sounds, 58–59
  "fighting purr," 59

R

reproductive disorders, hens
  egg binding, 142
  internal laying/egg yolk peritonitis, 142

mating injuries and, 141
symptoms of, 142
vent prolapse, 143
respiratory Infections, 138–39
roosts/roosting, 71–72, *71*, *72*
broad-breasted turkeys and, 72
construction, perfect, 51, *51*
wild instincts and, 50, *50*
roundworms, 140
run, secure, 128, *128*

S

scaly leg mites, 139–140
scratch, 90
seasonal care, 88–90, *88*, *90*
sellers, private, 96, *96*
sex of poults, determining, 131, *131*
shelter, 65–71, *118*. *See also* coop(s); housing needs, poults
daytime, 70, *70*
enclosed housing, 67
open-sided barn-type, 66, *66*, 68, *68*
simple, 69, *69*
successful, elements of, 69
temporary, 70–71
snood, *44*, 46, *46*
sounds/calls, 58–61
space requirements. *See* shelter
spit and drum sound, 58
spurs, *44*, 46, *46*
trimming, 141
strutting behavior, 48–49, *48*
summer care, 88–89, *88*
supplements, 84–85, *85*
poults and, 109
probiotics, herbs and, 109
suspicious behavior, 52, *52*

T

temperament, 25–26, *25, 26*
territorial behavior, 29
Thanksgiving turkeys, 13, 18, 19, 38, 66, 82
threatened breeds, 24
tom turkey, *21*
bad behavior of, 55–57
beard of, 44, *44*, 45, *45*
boss of/ dominance, 55
broody hens and, 119
coexisting, number of, 53–54, *54*
courtship display, 61–62, *61*, *62*
facial colors, 47, *47*
feeding time and, 28
heritage, weight at maturity, 37
photographing, 63, *63*
snood of, 46, *46*
sounds made by, 58–59
space requirements for multiple, 73
strutting behavior of, 48, *48*
territorial behavior, 29
weight at maturity, heritage, 37
tom-to-hen ratio, 54–55
treats
poults and, 110, 127
trying new, 84, *84*
turkey talk, 58–61

V

vent (cloaca), *44*, 46
vent prolapse, hen, 143
veterinary care, 135
vocalizations, 58–61

W

watch breeds, 24
water
apple cider vinegar added to, 109
clean source of, 83, *83*
summer and, 88–89, *88*
waterers, 104, 106–7
fount-style, 104
marbles or rocks in, 107, *107*
wattle, *44*, 46
wild instincts, 49–50
wild at heart story, 50
wing clipping, 77, *77*
winter care, 89–91, *90*
hardiness, winter, 89
snow turkeys story, 91, *91*
worms, intestinal, 140

Raise Poultry Successfully
with These Other Storey Books

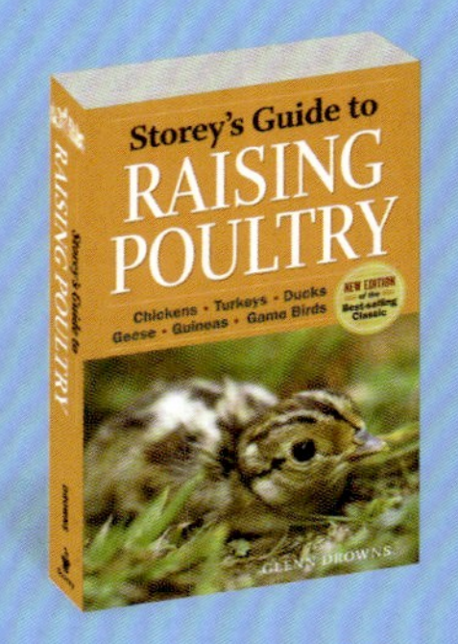

Storey's Guide to Raising Poultry, 4th Edition by Glenn Drowns
This guide has everything you need to know to humanely raise your own chickens, turkeys, waterfowl, and specialty birds, including doves and ostriches. Drowns provides expert advice on breed selection, housing, feeding, behavior, breeding, health care, and meat and egg processing.

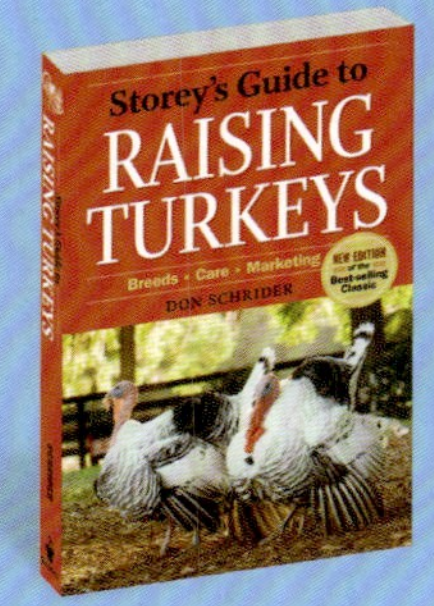

Storey's Guide to Raising Turkeys, 3rd Edition by Don Schrider
With Schrider's expert guidance, you'll learn how to raise healthy turkeys from start to finish. Includes tips on how to acquire organic certification, how to process meat and eggs, and how to successfully market your products.

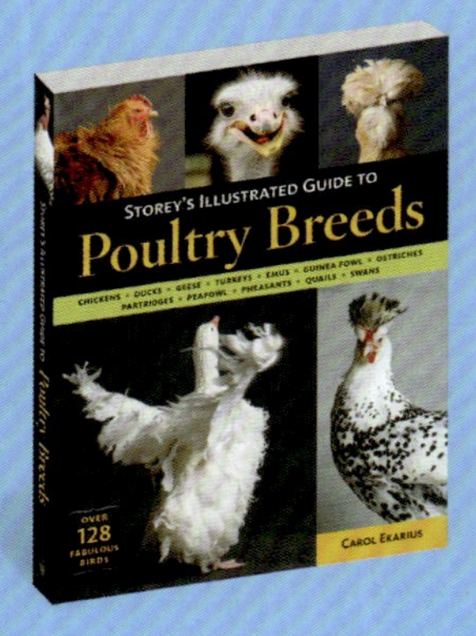

Storey's Illustrated Guide to Poultry Breeds by Carol Ekarius
From the Manx Rumpy to the Redcap, Ekarius provides 128 breed profiles, each with a brief history, a detailed description of identifying characteristics, and beautiful photography. Comprehensive and fun, this guide offers both essential information and an entertaining look at some of North America's quirkiest birds.

Join the conversation. Share your experience with this book, learn more about Storey Publishing's authors, and read original essays and book excerpts at storey.com. Look for our books wherever quality books are sold or call 800-441-5700.